Ensemble History Matching

Geir Evensen · Dean S. Oliver · Remus G. Hanea

Ensemble History Matching

Conditioning Reservoir Models on Dynamic Data

Springer

Geir Evensen
Data Assimilation and Optimization
NORCE Norwegian Research Center
Bergen, Norway

Dean S. Oliver
Energy and Technology
NORCE Norwegian Research Center
Bergen, Norway

Remus G. Hanea
Subsurface and Reservoir Technology
Equinor
Bergen, Norway

ISBN 978-3-031-99154-7 ISBN 978-3-031-99155-4 (eBook)
https://doi.org/10.1007/978-3-031-99155-4

This work was supported by NORCE Norwegian Research Center (NORCE Norwegian Research Center).

This Springer imprint is published by the registered company Springer Nature Switzerland AG
The registered company address is: Gewerbestrasse 11, 6330 Cham, Switzerland

Preface

This book aims to formulate the history-matching problem consistently and present state-of-the-art ensemble solution methods. By writing this monograph, we hope to contribute with a text that will help the practitioner in the field to understand the properties of ensemble methods better when used to history-match reservoir models. We also aim to provide a text that can serve an educational purpose for graduate students and researchers in petroleum, geothermal, and hydrological engineering and sciences. We will introduce and explain various algorithms used in data assimilation and parameter estimation, focusing on ensemble methods, particularly the most popular ones in the petroleum community. We will also discuss the challenges associated with these techniques, such as dealing with high-dimensional models, finite number of realizations, parameterization, and handling uncertainties in the observations and model parameters.

We have divided the book into three parts. Part I presents the mathematical formulation of the history-matching problem starting from Bayes' theorem. We derive three popular solution methods: the non-iterative Ensemble Smoother, the iterative ensemble-subspace version of the Ensemble Randomized Maximum Likelihood method, and the Ensemble Smoother with Multiple Data Assimilation. We also show how using the same numerical implementation for all these methods is possible. Additionally, we discuss an adaptive correlation-based localization method useful for petroleum applications with typically non-local measurements, and we present a version of the truncated pluri-Gaussian method that allows for updating categorical variables.

Part II introduces robust optimization and closed-loop reservoir management. We start by discussing the popular ensemble optimization method and continue with a presentation and discussion of a recent development, the mean-model bias correction method, which may be more efficient for some problems. The final chapter of Part II discusses closed-loop reservoir management, where we integrate history matching of past measurements with optimization for future production.

Part III discusses the practical use of ensemble methods for history-matching reservoir models. The purpose is to give an in-depth analysis of the methods' properties in practical use. We examine the importance of specifying the correct measurement error statistics, accounting for model errors, and how to select the observations on which we condition the model. A particular issue discussed in Sect. 13.6 considers the impact of redundancy in historical rate data, which is typically ignored in ensemble history matching and, in practice, leads to ensemble collapse and underestimation of the ensemble variance. In this discussion, we also consider the importance of using localization. We start by presenting a simple idealized case for illustrational purposes before we show some results from an extensive field application where Equinor uses ensemble history matching on the Troll field.

Finally, in the summary chapter, we briefly discuss implementing a history-matching application using open-source state-of-the-art ensemble tools, namely the Ensemble Reservoir Tool (ERT) provided by Equinor and the Python Ensemble Toolbox (PET) from NORCE Research. Furthermore, we reflect on the future development of ensemble methods for reservoir management within the petroleum industry.

We have based much of the data-assimilation theory and notation on the recent textbook on data assimilation by Evensen et al. (2022) but with a sole focus on history matching. Thus, we ignore the recursive nature of the sequential data-assimilation problem and focus only on the parameter estimation problem solved for petroleum applications.

At `https://github.com/geirev/Ensemble-History-Matching`, we will place history-matching examples and slides suitable for teaching. The repository also allows commenting and raising issues regarding the book's content. Thus, we encourage readers to report grammar, equations, or general discussion errors by submitting an *issue* to this repository. It is also possible to discuss the book's topics here.

During the work with this text, Geir Evensen received support from the Research Council of Norway and the companies Equinor, AkerBP, Wintershall–DEA through the Petromaks–2 research project REMEDY Grant 336240 and the Petro Center NCS2030 funded by the Research Council of Norway and several industry partners, Grant 331644. Dean Oliver was supported by REMEDY Grant 336240 while developing the mean model optimization method. The authors are grateful to Equinor and the Troll asset for allowing us to present the results in Chap. 14.

Bergen, Norway

Geir Evensen
Dean S. Oliver
Remus G. Hanea

Competing Interests The authors have no competing interests to declare that are relevant to the content of this manuscript.

Symbols

This list describes mathematical symbols used in the book.

Greek Symbols

α	Bias correction of in ensemble optimization Eq. (11.4)
α	Cummulative distribution in APS Eq. (8.1)
$\alpha(\mathbf{u}_i)$	Partial correction factor for $\mathbf{u}_i$ Eq. (11.10)
α^{pri}	Noninformative prior for b_j^{pri} Eq. (11.12)
α_i	MDA weights Eq. (3.18)
β_{ij}	Partial correction factor for $\mathbf{u}_i$ and $\boldsymbol{\theta}_j$ Eq. (11.8)
γ	Steplength in Gauss-Newton iterations Eq. (3.8)
$\boldsymbol{\eta}$	Innovation in iterative methods Eq. (3.13)
η	Steplength parameter in EnOpt Eq. (10.2)
$\boldsymbol{\theta}$	Vector of uncertain model parameters Eq. (2.2)
$\overline{\boldsymbol{\theta}}$	Mean vector of uncertain model parameters Eqs. (10.3) and (11.3)
$\widehat{\boldsymbol{\theta}}$	Surrogate mean vector of uncertain model parameters Eq. (11.1)
θ_λ	Box-Cox transformation Eq. (11.31)
λ	Levenberg-Marquart parameter Eq. (6.4)
λ	Power in Box-Cox transformation Eq. (11.31)
$\boldsymbol{\Lambda}$	Eigen values Eq. (6.21)
ν	The number of control variables in the optimization Eq. (11.8)
ξ	State-vector increment in iterative methods Eq. (3.14)
$\boldsymbol{\Pi}$	Projection subtracting the mean and scaling by $\sqrt{N-1}$ Eq. (5.2)
ρ_{trunc}	Truncation correlation for correlation based localization Eq. (7.4)
$\rho(i, m)$	Correlation been variable i and measurement m Eq. (7.5)
σ_β	Standard deviation of $\mathbf{C}_{bb}$ Eq. (11.30)
σ_b	Standard deviation for the prior b_j^{pri} Eq. (11.12)
σ_i	Standard deviation of measurement number i Eq. (2.1)

$\boldsymbol{\Sigma}$	Singular values of SVD Eq. (6.23)
$\boldsymbol{\Upsilon}$	Ensemble of model-predicted measurements Eq. (5.8)
$\boldsymbol{\Omega}$	Relation between initial and the ith iteration ensemble anomalies Eq. (6.11)
$\Omega(\mathbf{u}, \boldsymbol{\theta})$	Objective function of its arguments Eq. (10.1)

Math Symbols

$\mathbf{A}$	Ensemble of state-vector anomalies Eq. (5.3)
$\mathbf{b}_j$	Vector of partial correction factors Eq. (11.8)
b_j^{pri}	Prior for partial correction factor for $\mathbf{b}_j$ Eq. (11.11)
b	Variable computed in Eq. (7.8)
b_j	Partial correction factor for realization j given $\mathbf{u}$ Eq. (11.7)
$\mathbf{B}$	Matrix of partial correction factors Eq. (11.9)
$\mathbf{C}$	Matrix inverted in Eq. (5.16) defined in Eq. (5.17)
$\mathbf{C}_{bb}$	Covariance matrix of $\mathbf{b}_j$ Eq. (11.12)
$\mathbf{C}_{dd}$	Measurement error covariance matrix Eq. (2.4)
$\overline{\mathbf{C}}_{dd}$	Ensemble measurement error covariance matrix Eq. (5.7)
$\mathbf{C}_{uo}$	Covariance matrix of controls $\mathbf{u}$ with the resulting objective functions Eq. (10.3)
$\mathbf{C}_{uu}$	Covariance matrix of controls $\mathbf{u}$ Eq. (10.2)
$\overline{\mathbf{C}}_{yy}$	Ensemble covariance matrix of predicted measurements $\mathbf{y}$ Eq. (5.10)
$\mathbf{C}_{yz}$	Covariance matrix of $\mathbf{y}$ and $\mathbf{z}$ Eq. (4.1)
$\overline{\mathbf{C}}_{yz}$	Ensemble covariance matrix of $\mathbf{y}$ and $\mathbf{z}$ Eq. (5.11)
$\mathbf{C}_{zy}$	Covariance matrix of $\mathbf{z}$ and $\mathbf{y}$ Eq. (4.2)
$\overline{\mathbf{C}}_{yz}$	Ensemble covariance matrix of $\mathbf{z}$ and $\mathbf{y}$ Eq. (5.11)
$\mathbf{C}_{zz}$	State covariance matrix Eq. (2.3)
$\overline{\mathbf{C}}_{zz}$	Ensemble state covariance matrix Eq. (5.4)
d_i	Measurement number i Eq. (2.1)
$\mathbf{d}$	Measurements Sect. (1.4)
$\mathbf{d}_j$	Measurements realization number j Sect. (1.4)
$d_{\text{corr}}(i, m)$	Correlation distance used in correlation based localization Eq. (7.5)
d_{trunc}	Truncation distance for correlation based localization Eq. (7.6)
$\mathbf{D}$	Ensemble of perturbed measurements Eq. (5.5)
$E_{\max}$	Maximum measurement error inflation used in correlation based localization Eq. (7.8)
$E(i, m)$	Measurement error inflation used in correlation based localization Eq. (7.7)
$\mathbf{E}$	Ensemble of measurement perturbations Eq. (5.5)
$f(\cdot)$	The argument's probability density function (pdf) Sect. (2.2)
y_i	Predicted measurement number i Eq. (2.1)
$\mathbf{g}(\cdot)$	Vector function to map model inputs to predicted measurements Eq. (1.2)

$\mathbf{G}$	Tangent linear operator of $\mathbf{g}$, Eq. (2.10)
$\mathbf{G}_j$	Tangent linear operator of $\mathbf{g}$ for realization j, Eq. (3.4)
$\mathbf{G}$	Best-fit ensemble averaged model sensitivity Eq. (4.1)
$\overline{\mathbf{G}}$	Ensemble representation of best-fit ensemble averaged model sensitivity Eq. (6.6)
h	Truncation in APS Eq. (8.3)
$\mathbf{H}$	Measurement matrix Eq. (11.14)
$\mathbf{I}$	Identity matrix Eq. (11.20)
$\mathbf{I}_m$	Identity matrix of dimension m Eq. (6.25)
$\mathbf{I}_N$	Identity matrix of dimension N Eq. (5.2)
$\mathcal{J}(\cdot)$	Cost function of its argument Eq. (2.1)
m	Number of measurements
n_α	Number of ESMDA steps Eq. (3.18)
n	Dimension of the state vector
N_u	Dimension of control vectors Eq. (11.8)
$\mathcal{N}(\cdot, \cdot)$	Normal distribution Sect. (3.1)
N	Number of ensemble members or particles
P_1	Probability in APS Eq. (8.2)
P_2	Probability in APS Eq. (8.3)
$\mathbf{Q}$	Eigen vectors (6.21)
$\mathbf{S}$	Ensemble of possibly scaled predicted measurement anomalies Eq. (6.8)
$\mathcal{S}(\cdot)$	Cost function for bias correction estimation Eq. (11.16)
$\mathbf{T}$	Transition matrix for the ensemble update Eq. (7.1)
$\mathbf{u}$	Vector of model controls Eqs. (2.2), (10.1), and (11.1)
$\overline{\mathbf{u}}$	Mean vector of model controls Eq. (10.3)
$\mathbf{U}$	Left singular vectors of SVD Eq. (6.23)
$\mathbf{V}$	Right singular vectors of SVD Eq. (6.23)
$\mathbf{w}_j$	Transform weights for realization j Eq. (5.18)
$\mathbf{W}$	Ensemble of transform weights Eq. (6.12)
x	Truncation in APS Eq. (8.2)
y_i	Predicted measurement number i Eq. (2.1)
y	Gaussian random field Eq. (8.1)
$\mathbf{Y}$	Ensemble of model-predicted measurement anomalies Eq. (5.9)
$\mathbf{y}$	Model-predicted measurements Eq. (1.2)
$\mathbf{y}_j$	Model-predicted measurements realization number j Sect. (1.4)
$\mathbf{z}$	State vector Eq. (1.1)
$\mathbf{z}_j$	State vector realization number j Sect. (1.4)
$\delta\mathbf{z}$	State vector increment Eq. (3.11)
$\mathbf{Z}$	Ensemble of state vectors Eq. (5.1)

Other Symbols

$\Re$ The space of real numbers, often raised to a dimension like $\Re^{m \times n}$ to denote a matrix

1 Vector of length N with all elements equal to one Eq. (5.2)

1_j Vector of length j with all elements equal to one Eq. (11.24)

1_{N_u} Vector of length N_u with all elements equal to one Eq. (11.11)

Contents

List of Approximations

Part I
Solving the HM Problem

The book's Part I formulates the history-matching problem and its solution methods starting from Bayes' theorem.

Chapter 1
Introduction

Abstract This chapter introduces the book and cites other relevant monographs and review papers. It briefly discusses the formation of and some properties of petroleum reservoirs and provides some background on the reservoir-characterization problem. After that, it presents a review and timeline for using ensemble methods to history-match petroleum reservoirs. It also briefly introduces localization methods as such methods have become essential in ensemble-based history matching before outlining the book's content.

1.1 Reservoir Characterization

Oil reservoirs form through a geological process that spans millions of years. It begins in a depositional environment where organic matter, like microscopic algae and plankton, accumulates on the sea floor. The organic material becomes buried and subjected to heat and pressure; over time, it transforms into hydrocarbons like oil and gas. Subsequent tectonic and deformational processes, such as folding and faulting, can lead to oil migration from the source rock to a reservoir, enclosed by impermeable rocks trapping the hydrocarbons in a reservoir structure. This entire geological evolution takes millions of years. Oil reservoirs result from this lengthy and dynamic geological process, and their exploration and extraction require sophisticated methods to tap into these ancient accumulations of hydrocarbons.

Reservoir models are essential in petroleum engineering, as they allow us to simulate and understand the behavior of the subsurface reservoirs containing oil and gas. The building of reservoir models involves several critical steps. We gather geological data from well logs, seismic surveys, and core samples. These data provide information about the rock properties, such as porosity and permeability, and the geometry of the reservoir. Then, using these data and information from field analogs, we can build a three-dimensional geological reservoir model. Figure 1.1 shows a simplified cartoon illustrating a typical reservoir structure with traps formed by an anticline and an impermeable fault.

After the basic structure is in place, we need to analyze petrophysical properties and assign porosity, permeability, and rock types to the different reservoir layers, as

G. Evensen et al., *Ensemble History Matching*,
https://doi.org/10.1007/978-3-031-99155-4_1

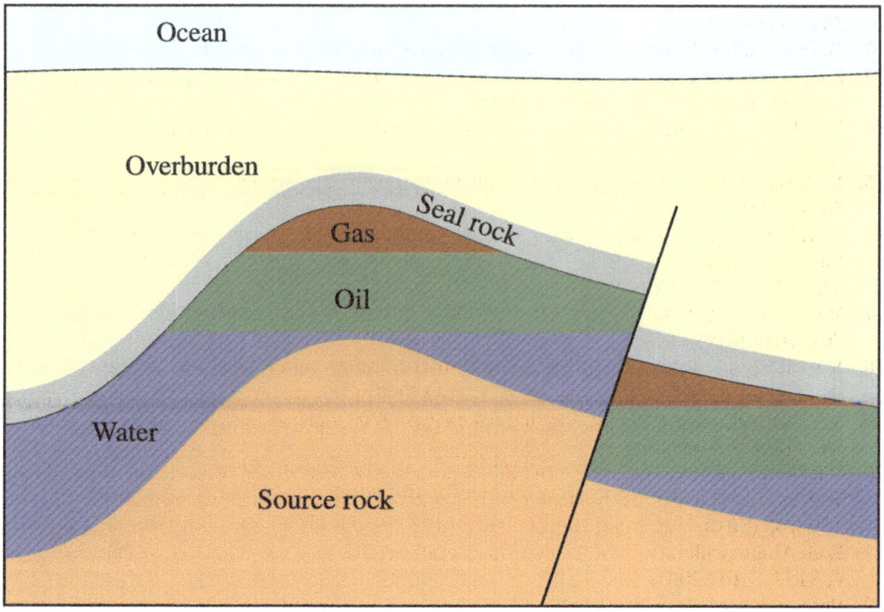

Fig. 1.1 Simplified cartoon showing a faulted sub-sea reservoir with hydrocarbon traps

these define how fluids are stored and flow in the reservoir. With this basic reservoir description, we continue by creating a suitable grid or mesh of the subsurface. During the property modeling, we assign properties such as porosity, permeability, and fluid saturations to each grid cell based on the geological model and petrophysical analysis.

After a fluid characterization, we can use reservoir simulators to solve mathematical equations that describe fluid flow in the reservoir. These simulations consider pressure, temperature, and fluid movement over time. It is then possible to run sensitivity analyses of the model to examine its response to changes in the input parameters, and we can assess the impact on the prediction uncertainties. We consider the range of possible outcomes in uncertainty quantification by considering uncertainties in geological and fluid parameters.

Significant uncertainties will be present in the reservoir description provided by the reservoir model. We build the structural framework from seismic data calibrated by a few well picks obtained from test wells. The wavelengths of the seismic signal limit the vertical resolution in the seismic data, and the uncertainty can be 10–30 meters at the typical reservoir depths. The horizontal resolution is typically much poorer than the vertical resolution. On the other hand, the reservoir's horizontal scales are often longer than the vertical scale, resulting from the initial smooth depositional environment. The structural uncertainties lead to uncertainties in the reservoir model's boundaries and grid definitions. Hence, we have an extremely challenging parameter estimation problem where we attempt to estimate the model grid, not only the model parameters.

After introducing the ES, different flavors of iterative ensemble smoother algorithms followed in the coming years. From 2011 to the present, the petroleum industry and academia have focused on three smoother algorithms, which we will consider in this book. In addition to ES, they are the Ensemble Smoother with Multiple Data Assimilation (ESMDA) proposed by Emerick and Reynolds (2012) and the ensemble randomized maximum likelihood method (EnRML) by Chen and Oliver (2012), Chen and Oliver (2013). We will use the recent ensemble subspace implementation of Evensen et al. (2019), Raanes et al. (2019) for the EnRML. The term "randomized maximum likelihood" is misleading. We solve for a randomized ensemble of maximum a posteriori solutions, as shown below.

1.4 The Ensemble History-Matching Approach

Based on the discussion in the previous section, we will now provide a descriptive and practical presentation of the ensemble history matching problem. We start by defining a "state vector," $\mathbf{z} \in \Re^n$, containing all the uncertain parameters we wish to improve the estimates of, see, e.g., Eq. (1.1) displayed in Fig. 1.3. The state vector contains parameters we assume to have significant uncertainty. In addition to purely static parameters, we have included the initial conditions for the oil-water and gas-oil contacts and the historical rate controls used to drive the model. This state vector serves as an input to a reservoir modeling workflow that we denote

$$\mathbf{y} = \mathbf{g}(\mathbf{z}). \tag{1.2}$$

Fig. 1.3 The state vector will typically include some or all of variables listed in Eq. (1.1)

$$\mathbf{z} = \left\{ \begin{array}{c} \text{Initial oil saturation} \\ \text{Initial gas saturation} \\ \text{Initial water saturation} \\ \text{Reservoir pressure} \\ \text{Historical oil production} \\ \text{Historical gas production} \\ \text{Historical water production} \\ \text{Porosity} \\ \text{Permeability} \\ \text{Structural surfaces} \\ \text{Fault multipliers} \\ \text{Facies description} \\ \text{Relative permeabilty} \\ \vdots \end{array} \right\}. \tag{1.1}$$

Here, the model operator $\mathbf{g}(\mathbf{z})$ is a composite function that can include, e.g., a geological model that creates a structural framework and populate it with reservoir properties to develop a reservoir-simulation model, followed by an execution of the reservoir model to simulate the predicted measurements $\mathbf{y} \in \Re^m$. We can then compare the predicted measurements to the observed measurements $\mathbf{d} \in \Re^m$, and the goal of the history matching method is to compute a new state vector that leads to a better fit to the observed measurements while preserving the geological consistency.

Given that we assign an uncertainty to the parameters in $\mathbf{z}$, we can represent this uncertainty by sampling an ensemble of model realizations $\mathbf{z}_j$ where the ensemble mean is the "best-guess" model, and the ensemble variance represents its uncertainty. Given an ensemble of model state vectors, we can simulate an ensemble of predicted measurements $\mathbf{y}_j$ using the model in Eq. (1.2). We then estimate the prediction uncertainty by propagating the initial and parameter uncertainty through the model system.

Ensemble methods work with such an ensemble of model states and use a form of regression to update the ensemble of model parameters such that the ensemble of model predictions becomes closer to the observed data, see Fig. 1.4. In this illustration, we have a monotonic response in the prediction from a change in the input parameter. We obtain an increase in the predicted value by increasing the parameter value. This monotonicity ensures a significant positive correlation between the simulated measurement and the input parameters, and it is this correlation that allows us to use ensemble methods for history-matching reservoir models. For example, increased permeability should enhance fluid flow and production, positively correlating permeability and production rates. In a realistic case, the situation becomes much more complicated as we can have thousands to millions of uncertain parameters and many measurements, including oil, gas, and water production and bottom hole pressures, from many wells. We often also have 4D-seismic data sets on which we wish to condition the reservoir description. A strength of the ensemble methods is their design, which allows them to handle this multi-variable and data situation and compute a consistent variance-minimizing estimate, considering all the prior information.

The relation between input parameters and predicted measurements is nonlinear, reducing their correlations. The nonlinear relationship is particular for the reservoir history-matching problem, where, e.g., the relation between porosity and a predicted oil rate comes from the simulation of the nonlinear reservoir model followed by a nonlinear well modeling. This situation is entirely different from many other data-assimilation applications where one directly observes a model's state variable and the correlation is identical to one at the measurement location. By examining the resulting correlation functions using increasingly larger ensemble sizes, we get an indication of how many realizations we need for computing consistent global updates.

In the case of a highly nonlinear model workflow, the correlations between the input parameters and the predicted measurements will weaken. Fortunately, experience has shown that, in most cases, we can use ensemble methods to improve the model parameterization in a history-matching workflow. Notably, introducing iterative ensemble smoothers has contributed to the success of ensemble history matching

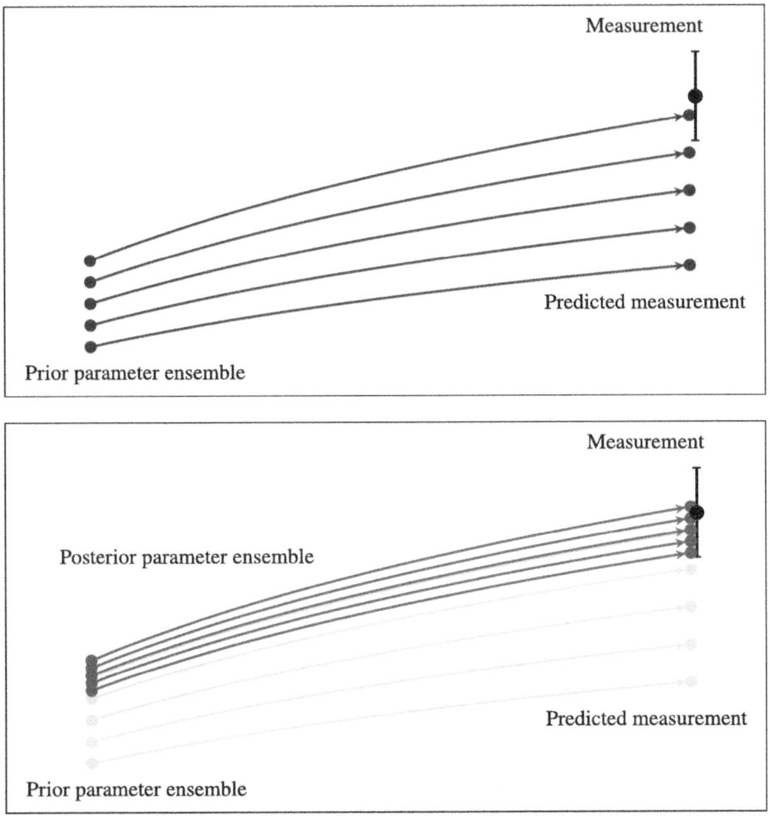

Fig. 1.4 The figure illustrates the linear regression update used to solve the parameter estimation problem. The upper frame displays the ensemble prediction of five realizations of a scalar parameter, which misses the measurement in blue. The lower frame shows how the updated parameter realizations lead to predictions better agreeing with the measurement (from Evensen et al. (2022))

since they allow for handling certain levels of nonlinearity much better than the original linear Ensemble Smoother (ES).

Before proceeding with the mathematical formulation and additional examples in the following chapters, we note that the ensemble methods can successfully estimate huge three-dimensional parameter fields because of the inherent correlation among the parameters. All the reservoir parameter fields are spatially correlated, significantly reducing the state vector's effective dimension. Additionally, we prescribe a prior model, ensuring we obtain a solution even when no data exists. Thus, the ensemble methods compute an estimate as an increment added to the prior model.

The ensemble method's central approximations are related to a model linearity assumption, which impacts several aspects of the technique and limits its use to weakly nonlinear problems and the use of a limited ensemble size due to computational constraints. We will discuss these assumptions in detail during the book's Part I.

One way to mitigate the impact of sampling errors, if one cannot just increase the ensemble size, is to apply a localization method. Localization methods reduce the effects of sampling errors and have enabled ensemble methods with small sample sizes of order $\mathcal{O}(10)$ realizations in several large scale realistic systems. The fol lowing section will briefly explain these methods as they are essential in most real applications, and in the remainder of the book, the introduction of practical adaptive correlation-based localization methods is central.

1.5 The Need for Localization in Ensemble Methods

Ensemble methods suffer from a limited ensemble size, which introduces sampling errors in the correlations used to update the model parameters. Increasing the ensemble size reduces the impact of sampling errors. But, how large must the ensemble size, N, be to ensure that the signal (physical correlation) to (sampling) noise ratio is sufficiently large to provide a consistent solution? As illustrated in Fig. 1.4, we use the correlations between uncertain parameters (like porosity and permeability fields) and the predicted measurements to compute linear updates to the input parameters.

In practical applications, computational resources and simulation time constraints will restrict the ensemble size we can afford to use. Thus, frequently, we need to reduce the ensemble size to well below the required number for computing consistent global updates. With smaller ensemble sizes, we introduce spurious updates in the posterior ensemble of model parameters when computing the global analysis. Interestingly, even with extensive spurious updates resulting from using a small ensemble size, we will often still obtain an excellent fit to the rate data on which we condition the model. It is, therefore, necessary to define what we mean by an excellent history-matching result. Many works measure the quality of the posterior parameter updates by the fit to the data conditioned on, without considering the updates to the uncertain model parameters. A better strategy is to *find the prior model parameters' minimal update that matches the data well.* This strategy requires eliminating significant spurious updates, and if we cannot run a sufficiently large ensemble, we must use localization.

When we update a parameter in a particular gridpoint, the localization methods ensure we only use information from "nearby" measurements. Hence, if no "nearby" measurements exist, we do not compute any update to the parameter. On the other hand, we have a certain flexibility in defining what we mean by "nearby." It can mean a physical or spatial distance in the numerical grid, leading to distance-based localization. We then introduce a physical truncation distance and ignore all measurements from wells located further away from the gridpoint we are updating. However, using a physical distance is only sometimes convenient in reservoir history matching. Estimating the correct truncation distance for different measurements is difficult, as long-range physical transients and the petrophysical fields introduce different scales in the model.

An alternative approach is to define the "nearby" measurements to include in a local update based on how strongly they correlate to the parameter we update. The measurement will not impact the update if the physical correlation is zero. Thus, we could use a strategy to truncate all measurements with a physical correlation less than a specific value. An issue with this strategy is that we need a large ensemble to determine the physical correlation, meaning we must resort to an approximate truncation scheme. We will introduce and demonstrate a correlation-based truncation in Chap. 7 and its use in the applications in Part III.

1.6 Outline of Book

The book's outline is as follows: In Part I, we will provide the detailed derivation of ensemble methods for history matching with emphasis on the Ensemble Smoother (ES), Ensemble Randomized Maximum Likelihood (EnRML), and Ensemble Smoother with Multiple Data Assimilation (ESMDA). Chapter 2 starts by reviewing and defining the history matching problem. The derivation starts from Bayes' theorem and introduces the assumption of Gaussian priors. This assumption leads to a quadratic cost function where the minimum defines the maximum a-posteriori solution of Bayes' theorem. We discuss how the cost function's gradient becomes central in the minimization methods.

Chapter 3 introduces the randomized maximum likelihood (RML) sampling, an approximate method for sampling the posterior Bayes'. In RML sampling, we solve many independent minimization problems like the one defined in Chap. 2. We show how the ensemble of cost functions leads to an ensemble of gradients, which results in an ensemble of minimizing solutions when set equal to zero. We can solve the gradient equation by introducing a linearization, which results in an ensemble of Kalman filter-type updates, or we can use Gauss-Newton iterations to solve the ensemble of minimization problems exactly. Finally, we explain the ESMDA approach as a tempering of the likelihood function, where one gradually introduces the data through a sequence of linear updates.

Chapter 4 explains how to replace the tangent linear and adjoint models with a linear regression expression based on covariance matrices, while in Chap. 5, we approximate all covariance matrices by an ensemble of realizations leading to a problem formulation in the ensemble subspace. In Chap. 6, we present the final subspace implementation of the EnRML algorithm from Evensen et al. (2019) and Raanes et al. (2019), and we explain how to use it to the solution for all of the ensemble methods.

The three last chapters of Part I cover some essential topics in reservoir history matching. In Chap. 7, we introduce and discuss the correlation-based adaptive localization method. After that, in Chap. 8, we present a truncated pluri-Gaussian method for history matching categorical variables, i.e., non-Gaussian facies variables. Finally, in Chap. 9, we discuss in more detail the impact of model nonlinearity on the ensemble methods and indicate which nonlinearity regimes we can expect the methods to work for.

In Part II, we will present so-called robust ensemble optimization methods, which have recently become popular in the petroleum community for optimizing, e.g., drainage strategies and field development planning. These methods benefit from the availability of a history-matched ensemble of reservoir models representing the reservoir uncertainty. By optimizing over an ensemble of reservoir models, it is possible to obtain a robust estimate for the future production accounting for the reservoir uncertainty. We will, in particular, consider the EnOpt method in Chap. 10, which uses a stochastic gradient to optimize the controls of an ensemble of reservoir models. However, we will also discuss some more recent and promising approaches in Chap. 11 that optimize using the mean model but introduce a bias correction.

Chapter 12 discusses a closed-loop-reservoir management workflow in a system that includes ensemble history matching and optimization. At a specific time, we use the history-matched ensemble of reservoir models as input to the ensemble optimization method to determine next year's drilling of wells and drainage strategy. Then, the following year, we will use newly acquired production data in another history-matching update before we repeat the optimization step. Thus, we have closed the loop. Finally, the ensemble of model predictions is input to a decision process. We will finalize Chap. 12 discussing how to use the ensemble predictions in a robust decision process where we account for the reservoir and prediction uncertainties.

In Part III, we will first discuss a simple reservoir case in Chap. 13, where we illustrate how to deal with measurement dependency and the redundancy of the information content in production data (Evensen 2021). We focus on the properties of the ensemble methods and the importance of ensemble size and localization. After this introductory example, we will move on to a large real case for the Troll reservoir on the Norwegian Continental Shelf in Chap. 14. This example includes a complete geological-to-simulation workflow and conditions the workflow on various data types, including 4D seismic data. We consider this Troll example as state-of-the-art for ensemble-based reservoir history matching. Finally, in the summary in Chap. 15, we will discuss some aspects of the ensemble history matching problem and challenges for the future.

Chapter 2
Formulating the History-Matching Problem

Abstract In this chapter, we will introduce and discuss the basic formulation of the history-matching problem. We start by presenting a brief historical summary of the development of the formulation of the history matching problem. Then, we introduce the fundamental assumption of a Gaussian prior and likelihood and connect the standard formulation of a quadratic cost function to Bayes' theorem. Finally, we discuss some possible methods for solving this problem.

2.1 Defining the History-Matching Problem

Historically, reservoir engineers have formulated the history-matching problem as *finding a set of model parameters in a simulation model leading to a simulation result that agrees with the observations*. The observations typically include oil, gas, and water production rates from several production wells. In addition, one commonly uses time series of measured bottom-hole pressures, water and gas injection rates from injection wells, and even 4D-seismic data.

In the first attempts of using *manual history matching* to fit a model to observed data, one would typically tune model parameters to adjust the simulated oil, gas, and water rates. The updated model parameters could include. e.g., multipliers that increase or decrease porosity and permeability in an area around a production well, fault transmissibilities, and relative permeability parameters. After a sequence of parameter adjustments, one would end up with a simulation model that fitted the observed rates better, but where the artificial updates of the petrophysical fields often resulted in a model inconsistent with the prescribed reservoir geology and likely with no predictability capabilities for new wells.

One could quantify the success of the manual history matching by defining a cost function for measuring the distance between a model prediction and a number, m, of observations, e.g.,

$$\mathcal{J}(\mathbf{z}) = \sum_{i=1}^{m} \frac{(d_i - y_i)^2}{\sigma_i^2} = \sum_{i=1}^{m} \frac{(d_i - g_i(\mathbf{z}))^2}{\sigma_i^2}. \tag{2.1}$$

G. Evensen et al., *Ensemble History Matching*,
https://doi.org/10.1007/978-3-031-99155-4_2

Here, the state vector, $\mathbf{z}$, contains the uncertain parameters to be adjusted, and we have defined the nonlinear reservoir model

$$\mathbf{y} = \mathbf{g}(\mathbf{z}) = \mathbf{g}(\boldsymbol{\theta}, \mathbf{u}), \tag{2.2}$$

which predicts the measured rates in $\mathbf{y}$. Thus, the nonlinear vector function, $\mathbf{g}(\mathbf{z})$, describes the forward simulation of the reservoir flow given the model parameters $\boldsymbol{\theta}$ and model controls $\mathbf{u}$ in $\mathbf{z}$ and then followed by the observation operator to obtain the predicted rates in $\mathbf{y}$. The model controls $\mathbf{u}$ typically represent model forcing, which in a reservoir simulation model comprises the imposed production and injection rates for the wells.

In the early history matching studies, the model parameters in $\mathbf{z}$ would include a few scalar parameters, e.g., permeability, porosity, and fault transmissibility multipliers. However, in modern history-matching applications, $\mathbf{z}$ can consist of the entire 3D fields of petrophysical variables, structural surfaces, and several scalar parameters related to, e.g., relative permeability, fault characteristics, and possibly parameters describing facies distributions. There is no limit to the uncertain parameters one may try to estimate. We will see below in Chap. 13 that we should also consider time-dependent model controls representing the historical production as uncertain model parameters. Thus, the history-matching problem is a high-dimensional nonlinear parameter estimation problem.

The choice of a quadratic cost function is natural for many reasons, but other alternatives exist. Here, σ_i is the standard deviation of the ith measurement. Note that this cost function only measures the total distance between all the included measurements and their model-predicted equivalents. As such, the cost function calculates a misfit between the observations and a prediction. However, the cost function has no statistical meaning in its current definition and form.

Following the initial attempts of manual history matching, there was a significant effort in formulating so-called assisted history-matching methods that would minimize the cost function in Eq. (2.1) using, e.g., gradient methods. However, some critical problems are associated with using the cost function in Eq. (2.1). The first and most fundamental one is that the formulation in Eq. (2.1) does not ensure a unique solution, even in the case of a linear reservoir model. Many parameter estimates may result in an exact fit to the observations and a value equal to zero for the cost function. To ensure a unique global minimum of $\mathcal{J}(\mathbf{z})$ from Eq. (2.1) in the linear case, we must have more independent information in the measurements than we have degrees of freedom in the parameters.

We will see later that a time series of production rates often contains highly redundant information. Thus, it is hard to know how many parameters we can estimate given a set of observations. We also risk finding noisy parameter estimates with large unphysical amplitudes. The data-assimilation community realized this issue early (e.g., Bennett 1992), and the apparent way to resolve it is to define a constraint for the prior parameter estimate. Thus, it is necessary to include another term in the cost function that penalizes the deviation from a first-guess or prior parameter value $\mathbf{z}^{\mathrm{f}}$. We use the superscript f for the first guess to be consistent with the notation in

Evensen et al. (2022). Hence, we add a prior penalty term to the cost function and write

$$\mathcal{J}(\mathbf{z}) = \frac{1}{2}\left(\mathbf{z} - \mathbf{z}^{\mathrm{f}}\right)^{\mathrm{T}} \mathbf{C}_{zz}^{-1} \left(\mathbf{z} - \mathbf{z}^{\mathrm{f}}\right) + \frac{1}{2} \sum_{i=1}^{m} \frac{(d_i - y_i)^2}{\sigma_i^2}. \tag{2.3}$$

The first term in Eq. (2.3) is quadratic and scaled by the error covariance of the parameters. The non-diagonal error covariance matrix $\mathbf{C}_{zz}$ ensures smooth updates of, e.g., the petrophysical field variables. The "prior" term aims to provide a unique minimizing solution for the linear case and to penalize large deviations from the first-guess values. In the nonlinear case, it is possible to have situations where different parameter values result in multiple global minima with the same cost-function value. Even without any observations, the cost function in Eq. (2.3) has a global minimum defined by the prior parameters. It also resolves the issue of excessive unphysical parameter updates as it penalizes deviations from the first-guess parameter values.

In the cost function from Eq. (2.3), it is also necessary to determine the relative weight between the prior and observation terms. Increasing the number of measurements will increase the contribution from the observation term in the cost function. In addition, with dependent information in the observations, they will impact the estimate too much. Some works have used a scaling factor for the measurement term to adjust the observations' impact relative to the prior term. However, it is possible to do better. We can define a more consistent cost function by accounting for measurement error correlations defined by a measurement error-covariance matrix, $\mathbf{C}_{dd}$, and write a more generic cost function as

Cost function

$$\mathcal{J}(\mathbf{z}) = \frac{1}{2}\left(\mathbf{z} - \mathbf{z}^{\mathrm{f}}\right)^{\mathrm{T}} \mathbf{C}_{zz}^{-1} \left(\mathbf{z} - \mathbf{z}^{\mathrm{f}}\right) + \frac{1}{2}\left(\mathbf{g}(\mathbf{z}) - \mathbf{d}\right)^{\mathrm{T}} \mathbf{C}_{dd}^{-1} \left(\mathbf{g}(\mathbf{z}) - \mathbf{d}\right). \tag{2.4}$$

Here, $\mathbf{z} \in \Re^n$ denotes the state vector of dimension n containing all the uncertain model parameters, and $\mathbf{C}_{zz} \in \Re^{n \times n}$ is its covariance matrix. The reservoir model $\mathbf{g}(\mathbf{z})$ maps the state vector to a set of predicted measurements $\mathbf{y} \in \Re^m$ that we can compare with the actual measurements in the vector $\mathbf{d} \in \Re^m$ with error covariance $\mathbf{C}_{dd} \in \Re^{m \times m}$.

The covariances $\mathbf{C}_{zz}$ and $\mathbf{C}_{dd}$ determine the relative weights between the prior parameters and the measurements. The covariance matrix, $\mathbf{C}_{zz}$, penalizes large deviations from the prior parameter values, and the role of the off-diagonal terms is to regularize and ensure smooth parameter updates respecting the parameters' physical scales. The measurement error covariance matrix, $\mathbf{C}_{dd}$, accounts for measurement dependency (measurements with correlated errors). Hence, dependent measurements will have less impact on the update than independent measurements.

The cost function in Eq. (2.4) is quadratic in both the prior and observation terms. We could define more elaborate cost functions, but the quadratic one has many valuable properties related to its interpretation and solution methods, as discussed below. Furthermore, the definition in Eq. (2.4) ensures a well-posed inverse problem. The minimizing solution is a parameter estimate close to the first guess values, resulting in predictions close to the observations. We can even find a minimizing solution without any available observations, in which case the result is just the prior estimate. Thus, the prior term is essential for making the inverse formulation well-posed.

2.2 Relation to Bayes' Theorem

Under the assumptions of Gaussian priors, minimizing the cost function in Eq. (2.4) is equivalent to maximizing Bayes' formula and thus solving for the *maximum a posteriori* (MAP) estimate. Hence, the formulation of the history-matching problem in terms of a cost function links directly to Bayes' theorem (or formula)

$$f(\mathbf{z}|\mathbf{d}) = \frac{f(\mathbf{d}|\mathbf{z})\, f(\mathbf{z})}{f(\mathbf{d})}, \tag{2.5}$$

where $f(\cdot)$ is the probability-density function of its argument.

Thus, we will now introduce the assumption of a Gaussian prior and likelihood function.[1]

Approximation 4 (Gaussian prior and likelihood.) We assume that the prior distributions of the state vector's components and observation errors are Gaussian distributed.

When assuming Gaussian distributions for the prior and likelihood we can write Bayes' theorem as

$$f(\mathbf{z}|\mathbf{d}) \propto \exp\big\{-\mathcal{J}(\mathbf{z})\big\}, \tag{2.6}$$

where the cost function, $\mathcal{J}(\mathbf{z})$, is the one in Eq. (2.4).

This direct link between Bayes' formula and the quadratic cost function is very fortunate as it allows us to define the history-matching problem in terms of basic principles. However, the minimizing solution's consistency depends on the proper specification of the prior and measurement error covariances. We also note that the Gaussian-priors assumption can lead to problems when dealing with model parameters with non-Gaussian distributions, such as facies parameters. We will discuss this issue in Chap. 8.

[1] We start the approximation numbering at four to ensure a one-to-one consistency with Evensen et al. (2022), where the three first approximations relate to the recursive data-assimilation problem's sequentiality and are irrelevant to the history-matching problem.

We will also point out that most state-of-the-art solution methods and software ignore the measurement error correlations and use a diagonal $\mathbf{C}_{dd}$ with the specified error variances on the diagonal. The apparent reason for neglecting the measurement error correlations is that they are often poorly known, and using a diagonal $\mathbf{C}_{dd}$ simplifies the update computation. However, as we will demonstrate in Chap. 13, this "simplification" may lead to severe problems with the measurement's impact becoming too excessive.

Another exciting aspect of the previous discussion is that we have used a "bottom-up" approach where we start with simple formulations and converge to a cost function in Eq. (2.4) that we can relate to Bayes' formula. The "bottom-up" development of the history-matching research has resulted in some peculiar issues. First, it is still usual to consider the measurement errors uncorrelated and use a diagonal measurement error covariance matrix. Second, we ignore essential model uncertainties related to uncertain model controls or forcings.

A top-down derivation of the history-matching problem starting from Bayes' theorem would inform us to include distributions for *all* uncertain parameters and their error statistics and then derive the corresponding cost function. This "top-down" approach immediately tells us to include a complete and representative error-covariance matrix for the measurements. In addition, we need to account for additional uncertain parameters not included in the standard formulation. We will revert to these points in Chap. 13.

Following the procedure outlined in this book will allow for consistent history matching with realistic error statistics for model parameters, controls, and predictions. An essential contribution of the current book demonstrates the importance of accounting for the uncertainty of the historical rates (or, in some cases, the historical bottom-hole pressures) when integrating the ensemble of model realizations. Also, we demonstrate the necessity of accounting for correlated measurement errors (or dependent data) in ensemble simulations and conditioning.

In Fig. 2.1, we illustrate how we solve the history matching problem by estimating new parameters and initial conditions that lead to a better data fit. The spread of the prior (blue) and posterior (green) bands indicate the prior and posterior uncertainty, while the thin lines could represent ensemble realizations.

2.3 Solution Methods

Traditional and straightforward assisted history-matching methods often used a gradient descent algorithm to minimize a version of the cost function in Eq. (2.4). A problem with gradient-based methods is that they search for the solution in the state space. For the reservoir models, the dimension of the space of uncertain parameters is vast, and the nonlinearity of the parameter estimation problem leads to a slow convergence. Thus, in practical applications, efficiency and convergence requirements often limit the number of parameters to order ten.

The gradient of the cost function in Eq. (2.4) is

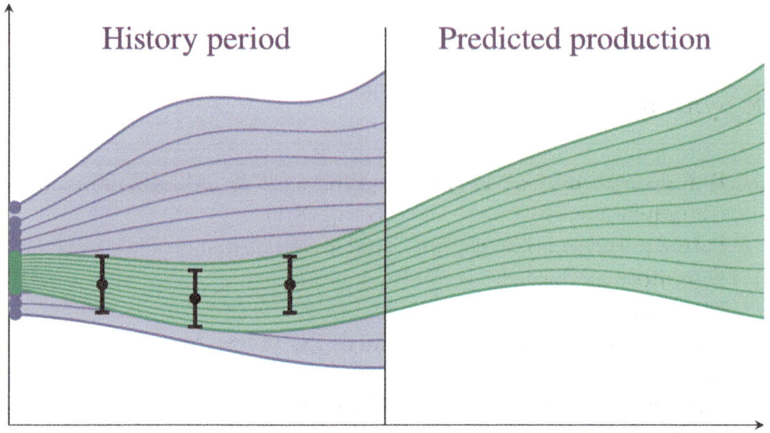

Fig. 2.1 The figure illustrates the smoother formulation assuming a perfect model solved using an ensemble approach. We define a time window in which we have observations and update the initial conditions and parameters of the ensemble realizations before rerunning the ensemble of updated models. We based this graphic on an illustration from the ECMWF Forecast User Guide https://confluence.ecmwf.int/display/FUG/Forecast+User+Guide

$$\nabla_{\mathbf{z}}\mathcal{J}(\mathbf{z}) = \mathbf{C}_{zz}^{-1}\big(\mathbf{z} - \mathbf{z}^{\mathrm{f}}\big) + \nabla_{\mathbf{z}}\mathbf{g}\big(\mathbf{z}\big)\mathbf{C}_{dd}^{-1}\big(\mathbf{g}(\mathbf{z}) - \mathbf{d}\big). \tag{2.7}$$

For large dimensions of the parameter space, the inverse of $\mathbf{C}_{zz}$ becomes challenging to compute, and similarly, for a large number of dependent measurements, the inverse of $\mathbf{C}_{dd}$ poses a computational problem. However, for the early cost-function formulations Eqs. (2.1) or (2.3), with a diagonal $\mathbf{C}_{dd}$ and a low dimensional $\mathcal{O}(10)$ parameter space, the adjoint method allowed for efficient evaluation of the gradient, using one backward integration of the adjoint tangent-linear model and one forward integration of the tangent-linear model. Alternatively, it was possible to evaluate the gradient by one forward model simulation per measurement as was done in Bissell et al. (1994). The problem is that these methods require knowledge of the model equations and their numerical implementation to develop the tangent-linear and adjoint models.

The use of gradient methods uncovers another issue with the cost function in Eq. (2.4), i.e., due to its nonlinearity, it may have multiple local and global minima, and in such cases, there is no guarantee of converging to a global minimum. Consequently, we must determine whether the converged solution is a global minimum or a local one. In an attempt to reduce the risk of converging to local minima, the focus shifted to Monte-Carlo sampling and evolutionary methods that, in the limit of infinite sample size, theoretically converge to the global minimum (Schulze-Riegert et al. 2002). Such evolutionary algorithms explore the solution space by random search and require numerous simulations to avoid getting trapped in local minima. However, their computational cost prevents them from finding the global minimum in practical applications, and the likelihood of convergence to a local minimum is significant. Contrary to the gradient methods, the sampling methods allow for the generation of uncertainty estimates.

Typical for these minimization methods is that they do not scale with the number of uncertain parameters defining the dimension of the problem. Although gradient methods can work well for many high-dimensional problems, the parameter estimation problem's nonlinearity renders both gradient methods and Monte Carlo sampling inefficient in high dimensions.

If we set the gradient in Eq. (2.7) equal to zero, we obtain an equation defining the cost function's minima:

Cost-function gradient set to zero

$$\mathbf{C}_{zz}^{-1}\left(\mathbf{z}-\mathbf{z}^{\mathrm{f}}\right)+\nabla_{\mathbf{z}}\mathbf{g}(\mathbf{z})\mathbf{C}_{dd}^{-1}\left(\mathbf{g}(\mathbf{z})-\mathbf{d}\right)=0. \tag{2.8}$$

This gradient equation does not have a closed-form solution due to the nonlinearity of the model $\mathbf{g}(\mathbf{z})$ and the evaluation of the model sensitivity $\nabla_{\mathbf{z}}\mathbf{g}$ at $\mathbf{z}$. However, a linearization approximation allows us to solve for $\mathbf{z}$:

Approximation 5 (Linearization) Linearize $\mathbf{g}(\mathbf{z})$ around the prior estimate $\mathbf{z}^{\mathrm{f}}$,

$$\mathbf{g}(\mathbf{z})\approx\mathbf{g}\left(\mathbf{z}^{\mathrm{f}}\right)+\mathbf{G}\left(\mathbf{z}-\mathbf{z}^{\mathrm{f}}\right), \tag{2.9}$$

and approximate the model gradient by evaluating it at the prior estimate

$$\mathbf{G}^{\mathrm{T}}=\nabla_{\mathbf{z}}\mathbf{g}(\mathbf{z})\big|_{\mathbf{z}=\mathbf{z}^{\mathrm{f}}}. \tag{2.10}$$

$\mathbf{G}$ is the tangent-linear operator of $\mathbf{g}(\mathbf{z})$ and $\mathbf{G}^{\mathrm{T}}$ is its adjoint. □

By applying the linearization in Approx. 5 Eq. (2.8), we can solve the linearized gradient equation for $\mathbf{z}$ to obtain

$$\mathbf{z}^{\mathrm{a}}=\mathbf{z}^{\mathrm{f}}+\mathbf{C}_{zz}\mathbf{G}^{\mathrm{T}}\left(\mathbf{G}\mathbf{C}_{zz}\mathbf{G}^{\mathrm{T}}+\mathbf{C}_{dd}\right)^{-1}\left(\mathbf{d}-\mathbf{g}\left(\mathbf{z}^{\mathrm{f}}\right)\right). \tag{2.11}$$

See the discussion in Sect. 3.2 for the derivation. For linear models, Eq. (2.11) provides the optimal or "best linear unbiased estimate" and forms the basis for the standard extended Kalman filter update. The equation requires the storage of the complete error covariance matrix $\mathbf{C}_{zz}$, which becomes prohibitive for large state vectors. Furthermore, we need to evaluate the tangent-linear model $\mathbf{G}$. In this book, we will not continue discussing the gradient or evolutionary algorithms for history matching. Instead, we will focus on ensemble methods for solving the gradient equation Eq. (2.8). We refer to the review by Oliver and Chen (2011), which provides a good account of many early works within manual and assisted history matching.

Rather than solving for the MAP estimate, ensemble methods attempt to sample approximately the Bayesian posterior and provide an estimate with associated uncertainty quantification. It is well-known that state-of-the-art ensemble methods would sample the Bayesian posterior precisely in the case of a linear model. In the case of modest nonlinearity, ensemble methods provide an approximate sampling as is demonstrated in Chap. 9.

Chapter 3
Randomized Maximum-Likelihood Sampling

Abstract We will now introduce the so-called randomized-maximum-likelihood sampling (RML), which provides a mean for sampling the posterior distribution in cases with modest nonlinearity. The method is a randomized a posteriori sampling, but we will use the standard notation common in the literature. The RML method assumes that we obtain approximate samples from the posterior distribution by minimizing an ensemble of specific cost functions. As this sampling is exact in the linear case, it is reasonable to assume that it provides approximate samples in the weakly nonlinear case. The RML sampling provides a fundamental basis for ensemble methods and allows us to connect these methods directly to Bayes' theorem.

3.1 Randomized Maximum-Likelihood Sampling

We start by introducing the following approximation:

Approximation 6 (Randomized Maximum-Likelihood sampling) approximately sample the posterior pdf with Gaussian priors by minimizing the ensemble of cost functions defined by Eq. (3.1) given below. □

RML is a method that approximately samples the Bayesian posterior. However, the name is inaccurate since the technique attempts to sample the posterior pdf, not the likelihood. Kitanidis (1995) and Oliver et al. (1996) appear to be the first papers introducing the method. In RML, we start by sampling an ensemble of prior state or parameter vectors $\mathbf{z}^{\mathrm{f}}_j \sim \mathcal{N}(\mathbf{z}^{\mathrm{f}}, \mathbf{C}_{zz})$ and an ensemble of measurement realizations $\mathbf{d}_j \sim \mathcal{N}(\mathbf{d}, \mathbf{C}_{dd})$, and then we minimize an ensemble of cost functions

G. Evensen et al., *Ensemble History Matching*,
https://doi.org/10.1007/978-3-031-99155-4_3

Ensemble of RML cost functions

$$\mathcal{J}(\mathbf{z}_j) = \frac{1}{2}\big(\mathbf{z}_j - \mathbf{z}^{\mathrm{f}}_j\big)^{\mathrm{T}}\mathbf{C}_{zz}^{-1}\big(\mathbf{z}_j - \mathbf{z}^{\mathrm{f}}_j\big) + \frac{1}{2}\big(\mathbf{g}(\mathbf{z}_j) - \mathbf{d}_j\big)^{\mathrm{T}}\mathbf{C}_{dd}^{-1}\big(\mathbf{g}(\mathbf{z}_j) - \mathbf{d}_j\big). \tag{3.1}$$

So, instead of minimizing one cost function in Eq. (2.4) to solve for the maximum *a posteriori* (MAP) estimate, we are minimizing an ensemble of independent cost functions in Eq. (3.1) to sample the Bayesian posterior. When the model $\mathbf{g}(\mathbf{z})$ is linear, RML samples exactly the posterior Bayesian pdf. In cases with modest nonlinearity, RML provides an approximate sampling of the Bayesian posterior. We have increased the problem of minimizing one cost function in Eq. (2.4) to minimize an ensemble of cost functions. We also note that the different cost function realizations and their associated gradient realizations are not connected, so we will sample independent realizations by minimizing the ensemble of cost functions.

For the minimization of Eq. (3.1), we set the ensemble of the cost function's gradients equal to zero

Ensemble of cost-function gradients set to zero

$$\mathbf{C}_{zz}^{-1}\big(\mathbf{z}_j - \mathbf{z}^{\mathrm{f}}_j\big) + \nabla_{\mathbf{z}}\mathbf{g}\big(\mathbf{z}_j\big)\mathbf{C}_{dd}^{-1}\big(\mathbf{g}(\mathbf{z}_j) - \mathbf{d}_j\big) = 0. \tag{3.2}$$

Thus, we must now solve an ensemble of independent gradient equations similar to Eq. (2.8).

3.2 Explicit Linear Solution

As discussed in the previous chapter, Eq. (3.2) has no explicit solution. However, by using the Approximation 5, it is possible to linearize the model operator around the prior state $\mathbf{z}^{\mathrm{f}}_j$ and write an approximate closed-form solution valid for small updates

Ensemble of Kalman-filter updates

$$\mathbf{z}^{\mathrm{a}}_j = \mathbf{z}^{\mathrm{f}}_j + \mathbf{C}_{zz}\mathbf{G}_j^{\mathrm{T}}\big(\mathbf{G}_j\mathbf{C}_{zz}\mathbf{G}_j^{\mathrm{T}} + \mathbf{C}_{dd}\big)^{-1}\big(\mathbf{d}_j - \mathbf{g}(\mathbf{z}^{\mathrm{f}}_j)\big). \tag{3.3}$$

We have defined the model's tangent linear operator as

$$\mathbf{G}_j^{\mathrm{T}} = \nabla_{\mathbf{z}}\mathbf{g}\big(\mathbf{z}_j\big). \tag{3.4}$$

For a detailed derivation of Eq. (3.3) using the Woodbury corollaries

Woodbury corollaries

$$\left(\mathbf{C}^{-1} + \mathbf{G}^{\mathrm{T}}\mathbf{D}^{-1}\mathbf{G}\right)^{-1} = \mathbf{C} - \mathbf{C}\mathbf{G}^{\mathrm{T}}(\mathbf{G}\mathbf{C}\mathbf{G}^{\mathrm{T}} + \mathbf{D})^{-1}\mathbf{G}\mathbf{C}, \tag{3.5}$$

$$(\mathbf{G}^{\mathrm{T}}\mathbf{D}^{-1}\mathbf{G} + \mathbf{C}^{-1})^{-1}\mathbf{G}^{\mathrm{T}}\mathbf{D}^{-1} = \mathbf{C}\mathbf{G}^{\mathrm{T}}\left(\mathbf{G}\mathbf{C}\mathbf{G}^{\mathrm{T}} + \mathbf{D}\right)^{-1}, \tag{3.6}$$

see the open-access book by Evensen et al. (2022, Sect. 6.5).

3.3 Solution by Gauss-Newton Iterations

However, we can avoid the linearization in Approximation 5 used to obtain the ensemble of updates in Eq. (3.3). By using iterative Gauss-Newton methods, we can minimize the ensemble of cost functions in Eq. (3.1) exactly. We then need the cost function's Hessian

$$\nabla_{\mathbf{z}}\nabla_{\mathbf{z}}\mathcal{J}(\mathbf{z}_j) \approx \mathbf{C}_{zz}^{-1} + \nabla_{\mathbf{z}}\mathbf{g}(\mathbf{z}_j)\mathbf{C}_{dd}^{-1}\left(\nabla_{\mathbf{z}}\mathbf{g}(\mathbf{z}_j)\right)^{\mathrm{T}}, \tag{3.7}$$

where we have neglected the terms involving second-order derivatives. The Gauss-Newton iteration becomes

Ensemble of Gauss-Newton iterations

$$\mathbf{z}_j^{i+1} = \mathbf{z}_j^i - \gamma\left(\mathbf{C}_{zz}^{-1} + \mathbf{G}_j^{i\,\mathrm{T}}\mathbf{C}_{dd}^{-1}\mathbf{G}_j^i\right)^{-1}\left(\mathbf{C}_{zz}^{-1}\left(\mathbf{z}_j^i - \mathbf{z}_j^{\mathrm{f}}\right) + \mathbf{G}_j^{i\,\mathrm{T}}\mathbf{C}_{dd}^{-1}\left(\mathbf{g}\left(\mathbf{z}_j^i\right) - \mathbf{d}_j\right)\right), \tag{3.8}$$

where γ is a step length, and we have defined the model's tangent-linear operator at iteration i and for ensemble member j as

$$\mathbf{G}_j^{i\,\mathrm{T}} = \nabla_{\mathbf{z}}\mathbf{g}\left(\mathbf{z}_j^i\right). \tag{3.9}$$

We note that both Eqs. (3.3) and (3.8) require the evaluation of the model's tangent linear operator at the prior or current model state. If the tangent-linear model is available, we can solve these equations directly using an ensemble of "adjoint" or "4DVar" methods. This approach is popular within the numerical weather prediction community, with a long tradition of developing and using the adjoint models in the data assimilation method. However, the model's tangent-linear operator is often unavailable, e.g., in commercial reservoir simulators where we only have access to an executable.

3.4 Incremental Form of Gauss-Newton Iterations

An alternative approach for minimizing the ensemble of cost functions in Eq. (3.1) is the incremental "ensemble 4DVar" formulation. The incremental 4DVar method is the standard approach in numerical weather prediction, but we have yet to see it used in petroleum applications. The idea is to solve recursively for a sequence of linear increments to the prior that leads to the posterior solution. For each increment, we must minimize a *linear quadratic cost function*, which is usually simpler than iterating directly on the nonlinear cost function.

Let's write Eq. (3.8) with $\gamma^i = 1$ as

$$\left(\mathbf{C}_{zz}^{-1} + \mathbf{G}_j^{i\,\mathrm{T}}\mathbf{C}_{dd}^{-1}\mathbf{G}_j^i\right)\left(\mathbf{z}_j^{i+1} - \mathbf{z}_j^i\right) = -\left(\mathbf{C}_{zz}^{-1}(\mathbf{z}_j^i - \mathbf{z}_j^{\mathrm{f}}) + \mathbf{G}_j^{i\,\mathrm{T}}\mathbf{C}_{dd}^{-1}\left(\mathbf{g}(\mathbf{z}_j^i) - \mathbf{d}_j\right)\right). \tag{3.10}$$

If we define

$$\delta\mathbf{z}_j = \mathbf{z}_j^{i+1} - \mathbf{z}_j^i, \tag{3.11}$$

Equation (3.10) also arises as the minimum of the following *quadratic* cost function for $\delta\mathbf{z}_j$

$$\begin{aligned}\mathcal{J}(\delta\mathbf{z}_j) = &\frac{1}{2}\left(\delta\mathbf{z}_j + \mathbf{z}_j^i - \mathbf{z}_j^f\right)^{\mathrm{T}}\mathbf{C}_{zz}^{-1}\left(\delta\mathbf{z}_j + \mathbf{z}_j^i - \mathbf{z}_j^f\right)\\ &+ \frac{1}{2}\left(\mathbf{G}_j^i\delta\mathbf{z}_j + \mathbf{g}(\mathbf{z}_j^i) - \mathbf{d}_j\right)^{\mathrm{T}}\mathbf{C}_{dd}^{-1}\left(\mathbf{G}_j^i\delta\mathbf{z}_j + \mathbf{g}(\mathbf{z}_j^i) - \mathbf{d}\right).\end{aligned} \tag{3.12}$$

This cost function linearizes the model and observation operators around the model trajectory for each Gauss-Newton iteration. Because $\delta\mathbf{z}_j$ is small, we can approximate $\mathbf{g}\left(\mathbf{z}_j^i + \delta\mathbf{z}_j\right) \approx \mathbf{g}\left(\mathbf{z}_j^i\right) + \mathbf{G}_j^i\delta\mathbf{z}_j$ in which $\mathbf{G}_j^i$ is the transpose of the gradient of $\mathbf{g}\left(\mathbf{z}_j^i\right)$. For convenience, we define the innovation vector

$$\boldsymbol{\eta}_j^i = \mathbf{d}_j - \mathbf{g}\left(\mathbf{z}_j^i\right), \tag{3.13}$$

and the residual

$$\boldsymbol{\xi}_j^i = \mathbf{z}_j^{\mathrm{f}} - \mathbf{z}_j^i. \tag{3.14}$$

With $\boldsymbol{\eta}_j$ and $\boldsymbol{\xi}_j$, we can now write the cost function in Eq. (3.12) for the increments $\delta\mathbf{z}_j$ as

Quadratic cost function for the increments

$$\mathcal{J}(\delta\mathbf{z}_j) = \frac{1}{2}\left(\delta\mathbf{z}_j - \boldsymbol{\xi}_j^i\right)^{\mathrm{T}}\mathbf{C}_{zz}^{-1}\left(\delta\mathbf{z}_j - \boldsymbol{\xi}_j^i\right) + \frac{1}{2}\left(\mathbf{G}_j^i\delta\mathbf{z}_j - \boldsymbol{\eta}_j^i\right)^{\mathrm{T}}\mathbf{C}_{dd}^{-1}\left(\mathbf{G}_j^i\delta\mathbf{z}_j - \boldsymbol{\eta}_j^i\right). \tag{3.15}$$

The solution for the increments becomes, from Eq. (3.10),

Solution for the increments

$$\left(\mathbf{C}_{zz}^{-1} + \mathbf{G}_j^{i\,\mathrm{T}} \mathbf{C}_{dd}^{-1} \mathbf{G}_j^i\right) \delta \mathbf{z}_j = \mathbf{C}_{zz}^{-1} \boldsymbol{\xi}_j^i + \mathbf{G}_j^{i\,\mathrm{T}} \mathbf{C}_{dd}^{-1} \boldsymbol{\eta}_j^i. \tag{3.16}$$

We can solve this linear set of equations for the increments iteratively, and we usually implement the approximate Hessian as a set of operations working on the vector $\delta\mathbf{z}_j$. Quasi-Newton and conjugate gradient methods are highly efficient for solving Eq. (3.16).

Thus, the incremental form of the Gauss-Newton method corresponds to an iterative scheme where we find the minimum of a quadratic cost function for $\delta\mathbf{z}_j$ in each iteration. After that, we update $\mathbf{z}_j^{i+1} = \mathbf{z}_j^i + \delta\mathbf{z}_j$ from (3.11), integrate the nonlinear model with the updated state vector, and recompute the variables $\boldsymbol{\eta}^i$ and $\boldsymbol{\xi}^i$ from Eqs. (3.13) and (3.14) before we solve Eq. (3.16) again. So, we have converted the non-quadratic and nonlinear minimization problem into a sequence of quadratic linear minimization problems. The only concern is whether the outer iteration in $\delta\mathbf{z}_j$ will converge.

3.5 Some Reflections

In the previous three sections, we have presented three approaches for minimizing the ensemble of RML cost functions. The first is an approximate method based on linearizations, while the latter are iterative methods that precisely solve for the ensemble of cost functions' minima. Notably, the iterative methods we've introduced are more than just theoretical constructs. They are practical tools that allow us to sample the Bayesian posterior with high accuracy, thanks to the Gaussian prior assumption and the RML sampling approximation. Note also that the solution for each ensemble member j is independent of the other ensemble members. It is, therefore, possible to solve for as many samples as we like, either sequentially on a uniprocessor computer or in parallel on a multiprocessor system.

The problem with the above solution methods is the adjoint model's numerical implementation requirement. The incremental approach is popular within the weather prediction community, where one has traditionally developed the adjoint model in conjunction with the nonlinear prediction model. As they previously solved for a single realization using the incremental 4DVar method, extending the system to solve multiple samples is straightforward and allows for better characterization of uncertainty. In this case, the meteorological community has named the method En4DVar (ensemble 4DVar).

Within the petroleum community, we typically run commercial reservoir simulators that do not provide the associated adjoint model. Thus, we must introduce further approximations to use the methods discussed in this chapter for these problems. We will discuss these approximations and the resulting methods in the following chapters. But first, in the following section, we will discuss an alternative and beneficial formulation for Bayes' theorem.

3.6 Multiple Data Assimilation

Multiple data assimilation (MDA) is an approach that reformulates Bayes' theorem to gradually introduce the information from measurements through an alternative representation of the likelihood function. Tempering of the likelihood function (Neal 1996) allows for rewriting it as a product of n_α inflated likelihoods.

Start by defining weights α_i, for $i = 1, \ldots, n_\alpha$, and the constraint

$$\sum_{i=1}^{n_\alpha} \frac{1}{\alpha_i} = 1. \tag{3.17}$$

It is then possible to rewrite Bayes' formula as

$$\begin{aligned} f\big(\mathbf{z}|\mathbf{d}\big) &\propto f\big(\mathbf{d}|\mathbf{g}(\mathbf{z})\big)\, f\big(\mathbf{z}\big) \\ &= f\big(\mathbf{d}|\mathbf{g}(\mathbf{z})\big)^{\left(\sum_{i=1}^{n_\alpha} \frac{1}{\alpha_i}\right)} f\big(\mathbf{z}\big) \\ &= f\big(\mathbf{d}|\mathbf{g}(\mathbf{z})\big)^{\frac{1}{\alpha_{n_\alpha}}} \cdots f\big(\mathbf{d} \mid \mathbf{g}(\mathbf{z})\big)^{\frac{1}{\alpha_2}} f\big(\mathbf{d} \mid \mathbf{g}(\mathbf{z})\big)^{\frac{1}{\alpha_1}} f\big(\mathbf{z}\big), \end{aligned} \tag{3.18}$$

leading to a procedure where we recursively introduce the measurement information by multiplying the prior with the inflated likelihood functions one by one. The first update step becomes

$$f_1\big(\mathbf{z}|\mathbf{d}\big) = f\big(\mathbf{d} \mid \mathbf{g}(\mathbf{z})\big)^{\frac{1}{\alpha_1}} f\big(\mathbf{z}\big), \tag{3.19}$$

followed by the second update step where $f_1\big(\mathbf{z}|\mathbf{d}\big)$ becomes the prior,

$$f_2\big(\mathbf{z}|\mathbf{d}\big) = f\big(\mathbf{d} \mid \mathbf{g}(\mathbf{z})\big)^{\frac{1}{\alpha_2}} f_1\big(\mathbf{z}|\mathbf{d}\big), \tag{3.20}$$

and so on for a total of n_α update steps. The MDA procedure solves the original Bayes' formula exactly and does not introduce any new approximations. We also note that the recursive update steps are independent of each other, a property that will become important in Sect. 6.4 when we derive the ESMDA method.

The choice of weights in the MDA factorization is irrelevant until we start computing approximate update steps, and we will revert to this issue in Sect. 6.4, where we introduce the Ensemble Smoother (ES) for the update steps.

With Gaussian priors, the likelihood function becomes

$$f\left(\mathbf{d}|\mathbf{g}(\mathbf{z})\right) \propto \exp - \left\{\frac{1}{2}\left(\mathbf{g}(\mathbf{z}_j) - \mathbf{d}_j\right)^{\mathrm{T}} \mathbf{C}_{dd}^{-1} \left(\mathbf{g}(\mathbf{z}_j) - \mathbf{d}_j\right)\right\}. \tag{3.21}$$

Thus, we have the inflated likelihood with the measurement error covariance multiplied by α,

$$f\left(\mathbf{d}|\mathbf{g}(\mathbf{z})\right)^{\frac{1}{\alpha}} \propto \exp - \left\{\frac{1}{2}\left(\mathbf{g}(\mathbf{z}_j) - \mathbf{d}_j\right)^{\mathrm{T}} (\alpha \mathbf{C}_{dd})^{-1} \left(\mathbf{g}(\mathbf{z}_j) - \mathbf{d}_j\right)\right\}. \tag{3.22}$$

It is, in principle, now possible to compute recursively n_α update steps that gradually introduce the observations using inflated observation errors. We could use the Gauss-Newton approach for each update step. However, since we now compute many minuscule updates, we can likely use the linear update Eq. (3.3) without introducing significant errors. This gradual introduction of the update reduces the impact of the linearization in Approximation 5 applied to obtain the update Eq. (3.3). However, since we need to compute the posterior distribution for each incremental update, as we need it for the next update step, we are not yet in a position to use the MDA approach. We will see in Chap. 6, how the ensemble formulation resolves this issue and renders the MDA approach beneficial for practical history-matching applications.

Chapter 4
Averaged Model Sensitivity

Abstract In this chapter, we will use the RML sampling strategy introduced in the previous chapter to introduce another simplification that eliminates the need for adjoint models. We will show that we can approximate the model sensitivity by a linear regression expressed by covariance matrices. Thus, we replace the adjoint and tangent linear models with a least squares best-fit linear regression approximation.

4.1 Linear Regression Approximation of the Model Sensitivity

Evensen (2018) used linearizations to show approximate connections $\mathbf{C}_{yz} \approx \mathbf{G}_j \mathbf{C}_{zz}$ and $\mathbf{C}_{yy} \approx \mathbf{G}_j \mathbf{C}_{zz} \mathbf{G}_j{}^{\mathrm{T}}$ that we can use to represent all appearances of the tangent-linear model, $\mathbf{G}_j$, in the update Eqs. (3.3) and (3.8), by covariance matrices. However, combining these two approximations introduces an inconsistency in the update equation (see Sect. 4.4, Evensen 2019).

Fortunately, we can take a slightly different approach in the derivation where we replace the tangent-linear operators defined in Eqs. (3.4) and (3.9) by a common averaged model sensitivity by applying the following approximation

Approximation 7 (Best-fit averaged model sensitivity) Interpret $\mathbf{G}_j$ in Eq. (3.3) and $\mathbf{G}_j^i$ in Eq. (3.8) as the sensitivity matrix in linear regression and represent them using the definition

$$\mathbf{G}_j \approx \mathbf{G} \triangleq \mathbf{C}_{yz} \mathbf{C}_{zz}^{-1}. \tag{4.1}$$

We then approximate the individual model sensitivities with a common averaged sensitivity for all realizations. □

Here, we have ignored the iteration index in Eq. (4.1), and $\mathbf{C}_{yz}$ is the covariance matrix between the predicted measurements $\mathbf{y}$ and the state vector $\mathbf{z}$. Note that we are

G. Evensen et al., *Ensemble History Matching*,
https://doi.org/10.1007/978-3-031-99155-4_4

now approximating the individual model sensitivities for each realization's gradient with an average best-fit one. Hence, we have changed the gradients to solve a different problem than we initially posed. So, the severity of the Approximation 7 depends on the ensemble spread and dissimilarities between the different model realizations. The benefit of introducing the approximation is that we replace the model tangent-linear and adjoint operators with covariance matrices.

4.2 Updated Solution Methods

Using Approximation 7, we can write the linear update Eq. (3.3) entirely in terms of the covariance matrices as

$$\mathbf{z}_j = \mathbf{z}_j^{\mathrm{f}} + \mathbf{C}_{zy}\left(\mathbf{C}_{yz}\mathbf{C}_{zz}^{-1}\mathbf{C}_{zy} + \mathbf{C}_{dd}\right)^{-1}\left(\mathbf{d}_j - \mathbf{g}\left(\mathbf{z}_j^{\mathrm{f}}\right)\right). \tag{4.2}$$

It is common to replace the term $\mathbf{G}\mathbf{C}_{zz}\mathbf{G}^{\mathrm{T}} = \mathbf{C}_{yz}\mathbf{C}_{zz}^{-1}\mathbf{C}_{zy}$ with $\mathbf{C}_{yy}$. However, most data-assimilation practitioners are unaware that this replacement introduces another approximation if $\mathbf{g}(\mathbf{z})$ is nonlinear as pointed out by Evensen (2019).

For the Gauss-Newton iteration in Eq. (3.8), the further derivation becomes slightly more intricate. We write the equation as

$$\begin{aligned}\mathbf{z}_j^{i+1} &= \mathbf{z}_j^i - \\ &\gamma\left(\left(\mathbf{C}_{zz}^i\right)^{-1} + \mathbf{G}^{i\,\mathrm{T}}\mathbf{C}_{dd}^{-1}\mathbf{G}^i\right)^{-1}\left(\mathbf{C}_{zz}^{-1}\left(\mathbf{z}_j^i - \mathbf{z}_j^{\mathrm{f}}\right) + \mathbf{G}^{i\,\mathrm{T}}\mathbf{C}_{dd}^{-1}\left(\mathbf{g}(\mathbf{z}_i) - \mathbf{d}_j\right)\right),\end{aligned} \tag{4.3}$$

where we, in the Hessian, for convenience, have evaluated the state covariance matrix at the current iterate (Chen and Oliver 2013) to ensure products $\mathbf{G}^i\mathbf{C}_{zz}^i$ with both the tangent-linear operator and the error covariances being at the ith iterate. Using the Woodbury corollaries, we can write Eq. (4.3) as

$$\begin{aligned}\mathbf{z}_j^{i+1} &= \mathbf{z}_j^i - \gamma\mathbf{C}_{zz}^i\mathbf{C}_{zz}^{-1}\left(\mathbf{z}_j^i - \mathbf{z}_j^{\mathrm{f}}\right) \\ &+ \gamma\mathbf{C}_{zz}^i\mathbf{G}^{i\,\mathrm{T}}\left(\mathbf{G}^i\mathbf{C}_{zz}^i\mathbf{G}^{i\,\mathrm{T}} + \mathbf{C}_{dd}\right)^{-1}\left(\mathbf{G}^i\mathbf{C}_{zz}^i\mathbf{C}_{zz}^{-1}\left(\mathbf{z}_j^i - \mathbf{z}_j^{\mathrm{f}}\right) - \left(\mathbf{g}(\mathbf{z}_j^i) - \mathbf{d}_j\right)\right).\end{aligned} \tag{4.4}$$

Now, using the averaged model sensitivity from the linear regression approximation in Eq. (4.1), we obtain the Gauss-Newton iteration on the form (Evensen 2019)

$$\begin{aligned}\mathbf{z}_j^{i+1} &= \mathbf{z}_j^i - \gamma\mathbf{C}_{zz}^i\mathbf{C}_{zz}^{-1}\left(\mathbf{z}_j^i - \mathbf{z}_j^{\mathrm{f}}\right) \\ &+ \gamma\mathbf{C}_{zy}^i\left(\mathbf{C}_{yz}^i\left(\mathbf{C}_{zz}^i\right)^{-1}\mathbf{C}_{zy}^i + \mathbf{C}_{dd}\right)^{-1}\left(\mathbf{C}_{yz}^i\mathbf{C}_{zz}^{-1}\left(\mathbf{z}_j^i - \mathbf{z}_j^{\mathrm{f}}\right) - \left(\mathbf{g}(\mathbf{z}_j^i) - \mathbf{d}_j\right)\right).\end{aligned} \tag{4.5}$$

Chen and Oliver (2013) discuss how to solve this equation, and they resolve the expensive matrix inversions by using a low-rank ensemble approximation and singular-value decompositions. They also introduce various approximate methods where, e.g., the product $\mathbf{C}^i_{zz}\mathbf{C}^{-1}_{zz} \approx \mathbf{I}$ and $\mathbf{C}^i_{yz}\mathbf{C}^{-1}_{zz}\left(\mathbf{z}^i_j - \mathbf{z}^{\mathrm{f}}_j\right) \approx 0$. However, in Chap. 6, we will re-derive the ensemble RML from a cost function formulated in the ensemble subspace and show how we effectively eliminate the costly inversions and end up with an efficient and correct algorithm.

Chapter 5
Ensemble Formulation

Abstract We will now introduce an ensemble representation or approximation for all the covariances appearing in Eqs. (4.2) and (4.5). This approximation leads to ensemble methods like EnKF and EnRML, which compute the updates using low-rank representations of the covariance matrices. We will see that the space spanned by the prior ensemble also contains the updated ensemble solution. Accordingly, we can simplify the derivation and formulation of the assimilation update equations by searching for the solution in the ensemble subspace defined by the prior ensemble.

5.1 Low-Rank Ensemble Representation

We now introduce the ensemble approximation of representing all covariance matrices by a finite number of realizations. For example, we will approximate the state covariance matrix $\mathbf{C}_{zz}$ using a finite number of state vector realizations. As the number of realizations is typically less than the dimension of the covariance matrix, this implies a low-rank approximation.

Approximation 8 (Ensemble approximation) A low-rank ensemble of model states with fewer realizations than the state dimension can approximately represent a covariance matrix. □

The RML formulation above has already introduced ensembles of realizations of the prior state vectors, $\mathbf{z}_j$, the predicted measurements, $\mathbf{y}_j$, and the perturbed measurements, $\mathbf{d}_j$. We can define low-rank approximations of the covariance matrices in Eqs. (4.2) and (4.5) from these ensembles.

We start by defining the prior ensemble of N model realizations, $\mathbf{z}_j \in \Re^n$, stored in the ensemble matrix $\mathbf{Z} \in \Re^{n \times N}$,

$$\mathbf{Z} = \left(\mathbf{z}_1, \mathbf{z}_2, \ldots, \mathbf{z}_N\right). \tag{5.1}$$

G. Evensen et al., *Ensemble History Matching*,
https://doi.org/10.1007/978-3-031-99155-4_5

Furthermore, we define the operator $\boldsymbol{\Pi} \in \Re^{N\times N}$ as

$$\boldsymbol{\Pi} = \left(\mathbf{I}_N - \frac{1}{N}\mathbf{1}\mathbf{1}^{\mathrm{T}}\right)/\sqrt{N-1}, \tag{5.2}$$

where $\mathbf{1} \in \Re^N$ is a vector with all elements equal to one and $\mathbf{I}_N$ is the N-dimensional identity matrix. Multiplying an ensemble matrix with $\boldsymbol{\Pi}$ subtracts the mean from the ensemble and scales the result by dividing it by $\sqrt{N-1}$. Note that, without the scaling, $\boldsymbol{\Pi}$ would define an orthogonal projection.

We can then define the zero-mean and scaled ensemble anomaly matrix as

$$\mathbf{A} = \mathbf{Z}\boldsymbol{\Pi}. \tag{5.3}$$

Thus, we approximate the state-covariance matrix by the ensemble covariance as

$$\mathbf{C}_{zz} \approx \overline{\mathbf{C}}_{zz} = \mathbf{A}\mathbf{A}^{\mathrm{T}}, \tag{5.4}$$

where the "overbar" denotes that we have a, possibly low-rank, *ensemble*-covariance matrix.

Correspondingly, we can define an ensemble of perturbed measurements, $\mathbf{D} \in \Re^{m\times N}$, when given the real measurement vector, $\mathbf{d} \in \Re^m$, as

$$\mathbf{D} = \mathbf{d}\mathbf{1}^{\mathrm{T}} + \sqrt{N-1}\mathbf{E}, \tag{5.5}$$

where $\mathbf{E} \in \Re^{m\times N}$ is the centered measurement-perturbation matrix with columns sampled from $\mathcal{N}(0, \mathbf{C}_{dd})$ and divided by $\sqrt{N-1}$. This definition also implies that, similarly to Eq. (5.3), we have

$$\mathbf{E} = \mathbf{D}\boldsymbol{\Pi}. \tag{5.6}$$

Hence, we represent the measurement error-covariance matrix by the ensemble-covariance matrix for the measurement perturbations as

$$\mathbf{C}_{dd} \approx \overline{\mathbf{C}}_{dd} = \mathbf{E}\mathbf{E}^{\mathrm{T}}. \tag{5.7}$$

From an ensemble of model integrations we compute the ensemble of predicted measurements

$$\boldsymbol{\Upsilon} = \mathbf{g}(\mathbf{Z}), \tag{5.8}$$

with scaled anomalies

$$\mathbf{Y} = \boldsymbol{\Upsilon}\boldsymbol{\Pi}, \tag{5.9}$$

where we have multiplied the model prediction by $\boldsymbol{\Pi}$ to subtract the ensemble mean and divide the resulting anomalies by $\sqrt{N-1}$.

Thus, we can also approximate $\mathbf{C}_{yy}$ and $\mathbf{C}_{yz}$ as

$$\mathbf{C}_{yy} \approx \overline{\mathbf{C}}_{yy} = \mathbf{Y}\mathbf{Y}^{\mathrm{T}}, \tag{5.10}$$

and

$$\mathbf{C}_{yz} \approx \overline{\mathbf{C}}_{yz} = \mathbf{Y}\mathbf{A}^{\mathrm{T}}. \tag{5.11}$$

5.2 Solutions Using Ensemble Covariances

A standard approach to derive the ensemble methods is to insert the ensemble covariance matrices into the "extended Kalman filter update" in Eq. (4.2) to obtain the EnKF or into the Gauss-Newton update Eq. (4.5) to get the EnRML method.

For EnKF or ES, we compute the minimizing solution of the ensemble of cost functions in Eq. (3.1) from Eq. (4.2), but using the ensemble of realizations to represent the error covariance matrices. Thus, we can write Eq. (4.2) as

$$\mathbf{z}_j^{\mathrm{a}} = \mathbf{z}_j^{\mathrm{f}} + \overline{\mathbf{C}}_{zy}\left(\overline{\mathbf{C}}_{yz}\overline{\mathbf{C}}_{zz}^{\dagger}\overline{\mathbf{C}}_{zy} + \overline{\mathbf{C}}_{dd}\right)^{-1}\left(\mathbf{d}_j - \mathbf{g}(\mathbf{z}_j^{\mathrm{f}})\right), \tag{5.12}$$

where the "dagger" symbol, †, denotes pseudo inverse. Equation (5.12) represents the EnKF update equation of Evensen (1994) with the perturbed observations proposed by Burgers et al. (1998) and is identical to the ES update in the parameter estimation problem discussed here.

Starting with Eq. (5.12), we insert the ensemble covariances using the representations in Eqs. (5.4), (5.7), and (5.11), and we obtain for the ensemble update,

$$\begin{aligned} \mathbf{Z}^{\mathrm{a}} &= \mathbf{Z}^{\mathrm{f}} + \mathbf{A}\mathbf{Y}^{\mathrm{T}}\left(\mathbf{Y}\mathbf{A}^{\mathrm{T}}(\mathbf{A}\mathbf{A}^{\mathrm{T}})^{\dagger}\mathbf{A}\mathbf{Y}^{\mathrm{T}} + \mathbf{E}\mathbf{E}^{\mathrm{T}}\right)^{-1}\left(\mathbf{D} - \mathbf{g}(\mathbf{Z}^{\mathrm{f}})\right) \\ &= \mathbf{Z}^{\mathrm{f}} + \mathbf{A}\mathbf{Y}^{\mathrm{T}}\left(\mathbf{Y}\mathbf{A}^{\dagger}\mathbf{A}\mathbf{Y}^{\mathrm{T}} + \mathbf{E}\mathbf{E}^{\mathrm{T}}\right)^{-1}\left(\mathbf{D} - \mathbf{g}(\mathbf{Z}^{\mathrm{f}})\right), \end{aligned} \tag{5.13}$$

where we have used the identity $\mathbf{A}^{\dagger} = \mathbf{A}^{\mathrm{T}}(\mathbf{A}\mathbf{A}^{\mathrm{T}})^{\dagger}$.

Corollary: It is straightforward to show that

$$\begin{aligned} \overline{\mathbf{C}}_{yz}\overline{\mathbf{C}}_{zz}^{\dagger}\overline{\mathbf{C}}_{zy} &= \mathbf{Y}\mathbf{A}^{\mathrm{T}}\left(\mathbf{A}\mathbf{A}^{\mathrm{T}}\right)^{\dagger}\mathbf{A}\mathbf{Y}^{\mathrm{T}} \\ &= \mathbf{Y}\left(\mathbf{A}^{\dagger}\mathbf{A}\right)\left(\mathbf{Y}\left(\mathbf{A}^{\dagger}\mathbf{A}\right)\right)^{\mathrm{T}} \\ &= \mathbf{Y}\mathbf{Y}^{\mathrm{T}} = \overline{\mathbf{C}}_{yy} \qquad \text{for } n \geq N-1, \end{aligned} \tag{5.14}$$

by using the following result from Sakov et al. (2012),

$$\mathbf{A}^{\dagger}\mathbf{A} = \mathbf{I}_N - \frac{1}{N}\mathbf{1}\mathbf{1}^{\mathrm{T}} \qquad \text{for } n \geq N-1. \tag{5.15}$$

However, only in the low-rank case, when $n \geq N - 1$, is $\mathbf{A}^{\dagger}\mathbf{A}$ a projection that removes the ensemble mean as defined in Eq. (5.15). Since in Eq. (5.14), the mean of $\mathbf{Y}$ is already zero by definition in Eq. (5.9), the additional multiplication with $\mathbf{A}^{\dagger}\mathbf{A}$ has no effect when $n \geq N - 1$.

Evensen (2003) introduced the formulation of the update Eq. (5.13) in terms of the ensemble-anomaly matrices. He further showed that this formulation leads to an update written as a linear combination of the prior ensemble, i.e., we can write

$$\begin{aligned} \mathbf{Z}^{\mathrm{a}} &= \mathbf{Z}^{\mathrm{f}} + \mathbf{A}\mathbf{Y}^{\mathrm{T}}\mathbf{C}^{-1}\big(\mathbf{D} - \mathbf{g}(\mathbf{Z}^{\mathrm{f}})\big) \\ &= \mathbf{Z}^{\mathrm{f}}\Big(\mathbf{I}_N + \boldsymbol{\Pi}\mathbf{Y}^{\mathrm{T}}\mathbf{C}^{-1}\big(\mathbf{D} - \mathbf{g}(\mathbf{Z}^{\mathrm{f}})\big)\Big), \end{aligned} \tag{5.16}$$

where we have defined the matrix $\mathbf{C} \in \Re^{m \times m}$ as

$$\mathbf{C} = \mathbf{Y}\mathbf{A}^{\dagger}\mathbf{A}\mathbf{Y}^{\mathrm{T}} + \mathbf{E}\mathbf{E}^{\mathrm{T}}. \tag{5.17}$$

We will discuss the inversion of $\mathbf{C}$ in Sect. 6.5 below while noting that a stable inversion is critical for the ensemble methods. From Eq. (5.16), it is clear that the ensemble subspace defined by the prior realizations also contains the updated ensemble's realizations.

Similarly, we can follow the procedure used by Chen and Oliver (2013) and introduce the ensemble matrices to the GN-update in Eq. (4.5). However, this procedure leads to a rather elaborate equation as we evaluate the covariance matrices at iteration zero and i, and we need to compute the inverse or pseudo-inverse of $\mathbf{C}_{zz}$. For now, we note that Eq. (4.5), when we insert the ensemble covariances from Eq. (5.4), implies that also the GN-iterations will lead to a posterior ensemble retained in the prior ensemble subspace. In the following section, we will use this finding and reformulate the cost function to search for the solution in the ensemble subspace. This procedure allows us to solve for the update in a low-dimensional ensemble subspace of dimension equal to the ensemble size.

5.3 Cost Function in the Ensemble Subspace

From the discussion in the previous section, we see that the leftmost matrix in the gradient is the ensemble anomaly matrix. Thus, we will search for the solution in the ensemble subspace spanned by the prior ensemble by assuming that an updated ensemble realization, $\mathbf{z}^{\mathrm{a}}_j$, is equal to the prior realization, $\mathbf{z}^{\mathrm{f}}_j$, plus a linear combination of the ensemble anomalies,

$$\mathbf{z}^{\mathrm{a}}_j = \mathbf{z}^{\mathrm{f}}_j + \mathbf{A}\mathbf{w}_j. \tag{5.18}$$

Following Hunt et al. (2007), we write the cost function (3.1) in terms of $\mathbf{w}_j$ as

Cost function in ensemble subspace

$$\mathcal{J}(\mathbf{w}_j) = \frac{1}{2}\mathbf{w}_j^{\mathrm{T}}\mathbf{w}_j + \frac{1}{2}\Big(\mathbf{g}\big(\mathbf{z}_j^{\mathrm{f}} + \mathbf{A}\mathbf{w}_j\big) - \mathbf{d}_j\Big)^{\mathrm{T}}\overline{\mathbf{C}}_{dd}^{-1}\Big(\mathbf{g}\big(\mathbf{z}_j^{\mathrm{f}} + \mathbf{A}\mathbf{w}_j\big) - \mathbf{d}_j\Big), \tag{5.19}$$

where we have used that

$$\begin{aligned}
\mathbf{w}_j^{\mathrm{T}}\mathbf{A}^{\mathrm{T}}\mathbf{C}_{zz}^{-1}\mathbf{A}\mathbf{w}_j &\approx \mathbf{w}_j^{\mathrm{T}}\mathbf{A}^{\mathrm{T}}\overline{\mathbf{C}}_{zz}^{-1}\mathbf{A}\mathbf{w}_j \\
&= \mathbf{w}_j^{\mathrm{T}}\mathbf{A}^{\mathrm{T}}\big(\mathbf{A}\mathbf{A}^{\mathrm{T}}\big)^{\dagger}\mathbf{A}\mathbf{w}_j \\
&= \mathbf{w}_j^{\mathrm{T}}\big(\mathbf{A}^{\dagger}\mathbf{A}\big)\mathbf{w}_j \\
&= \mathbf{w}_j^{\mathrm{T}}\big(\mathbf{A}^{\dagger}\mathbf{A}\big)\big(\mathbf{A}^{\dagger}\mathbf{A}\big)\mathbf{w}_j \\
&= \widetilde{\mathbf{w}}_j^{\mathrm{T}}\widetilde{\mathbf{w}}_j.
\end{aligned} \tag{5.20}$$

in which we defined $\widetilde{\mathbf{w}}_j = \big(\mathbf{A}^{\dagger}\mathbf{A}\big)\mathbf{w}_j$. The expression $\mathbf{A}^{\dagger}\mathbf{A}$ is the orthogonal projection onto the range of $\mathbf{A}^{\mathrm{T}}$. Also, we have the projection property $\big(\mathbf{A}^{\dagger}\mathbf{A}\big)\big(\mathbf{A}^{\dagger}\mathbf{A}\big) = \mathbf{A}^{\dagger}\mathbf{A}$. Thus, $\widetilde{\mathbf{w}}_j$ is just the projection of $\mathbf{w}$ onto ensemble perturbation space, and from

$$\mathbf{A}\mathbf{w}_j = \mathbf{A}\big(\mathbf{A}^{\dagger}\mathbf{A}\big)\mathbf{w}_j = \mathbf{A}\widetilde{\mathbf{w}}_j, \tag{5.21}$$

we see that it does not matter whether we solve for $\mathbf{w}_j$ or $\widetilde{\mathbf{w}}_j$.

Note that we have used an ensemble representation for the measurement-error-covariance matrix. We could have retained the complete $\mathbf{C}_{dd}$, but this matrix is too large for practical computations in many cases, and a diagonal matrix commonly approximates it. Thus, one typically neglects all measurement error correlations, which can have dire consequences. Evensen (2021) proposed using the ensemble representation in Eq. (5.7) but with an increased ensemble size to mitigate additional sampling errors (see Sect. 6.5).

Minimizing the cost functions in Eq. (5.19) implies solving for the minima of the original cost functions in Eq. (3.1), but restricted to the ensemble subspace and with $\overline{\mathbf{C}}_{zz}$ in place of $\mathbf{C}_{zz}$, as explained by Bocquet et al. (2015). The ensemble of cost functions in Eq. (5.19) does not refer to the high-dimensional state-covariance matrix, $\mathbf{C}_{zz}$. We searched for the solution in the state space in the original formulation in Chap. 3. We now have a more straightforward problem searching for the ensemble subspace solution. Thus, we solve for the N vectors $\mathbf{w}_j \in \Re^N$, one for each realization.

Chapter 6
Subspace EnRML

Abstract This chapter derives Gauss-Newton and Levenberg-Marquardt methods for RML sampling that searches for the solution in the ensemble subspace. The approach does not introduce any further assumptions since ensemble methods implicitly confine the solution to the space spanned by the prior ensemble. The advantage of the ensemble subspace formulation is that it avoids the inversion of huge low-rank covariance matrices and solves the EnRML formulation precisely without introducing any further approximations. The subspace EnRML algorithm can also compute the posterior ensemble subspace solution for the ES and ESMDA methods.

6.1 Levenberg-Marquardt in the Ensemble Subspace

Evensen et al. (2019) formulated the Gauss-Newton method for minimizing the cost function in Eq. (5.19). Here, we will take a slightly different approach and derive the Levenberg-Marquart method instead. Levenberg-Marquardt modifies the Hessian somewhat by adding a term $\lambda \mathbf{I}_N$ and can stabilize the iteration in some cases.

The Jacobian (gradient) of the cost function $\nabla_{\mathbf{w}} \mathcal{J}(\mathbf{w}_j) \in \mathcal{R}^{N\times 1}$ is

$$\nabla_{\mathbf{w}} \mathcal{J}(\mathbf{w}_j) = \mathbf{w}_j + \left(\mathbf{G}_j \mathbf{A}\right)^{\mathrm{T}} \overline{\mathbf{C}}_{dd}^{-1} \left(\mathbf{g}\left(\mathbf{z}_j^{\mathrm{f}} + \mathbf{A}\mathbf{w}_j\right) - \mathbf{d}_j\right), \tag{6.1}$$

and an approximate Hessian (gradient of the Jacobian) $\nabla_{\mathbf{w}}\nabla_{\mathbf{w}} \mathcal{J}(\mathbf{W}) \in \mathcal{R}^{N\times N}$ becomes

$$\nabla_{\mathbf{w}}\nabla_{\mathbf{w}} \mathcal{J}\left(\mathbf{w}_j\right) \approx \mathbf{I}_N + \left(\mathbf{G}_j \mathbf{A}\right)^{\mathrm{T}} \overline{\mathbf{C}}_{dd}^{-1} \left(\mathbf{G}_j \mathbf{A}\right). \tag{6.2}$$

We have defined the tangent-linear model

$$\mathbf{G}_j = \left(\nabla_{\mathbf{w}} \mathbf{g}|_{\mathbf{z}_j^{\mathrm{f}} + \mathbf{A}\mathbf{w}_j}\right)^{\mathrm{T}} \in \mathcal{R}^{m\times n}, \tag{6.3}$$

and in the Hessian, we have neglected the second-order derivatives.

The iterative Levenberg-Marquart scheme for minimizing the cost function in Eq. (5.19) becomes

G. Evensen et al., *Ensemble History Matching*,
https://doi.org/10.1007/978-3-031-99155-4_6

$$\mathbf{w}_j^{i+1} = \mathbf{w}_j^i - \gamma^i \Big((1+\lambda^i)\mathbf{I}_N + \big(\mathbf{G}_j^i\mathbf{A}\big)^\mathrm{T} \overline{\mathbf{C}}_{dd}^{-1} \big(\mathbf{G}_j^i\mathbf{A}\big) \Big)^{-1} \\ \times \Big(\mathbf{w}_j^i + \big(\mathbf{G}_j^i\mathbf{A}\big)^\mathrm{T} \overline{\mathbf{C}}_{dd}^{-1} \big(\mathbf{g}\big(\mathbf{z}_j^\mathrm{f} + \mathbf{A}\mathbf{w}_j^i\big) - \mathbf{d}_j \big) \Big), \tag{6.4}$$

where we have now introduced λ^i as an additional parameter. Thus, it is clear that when λ is large, the Hessian approaches $\lambda^i \mathbf{I}_N$, and we obtain the gradient descent method with a small step length of γ^i/λ^i. On the other hand, when $\lambda = 0$, we revert to the Gauss-Newton method. Levenberg-Marquart will usually converge slower than the standard Gauss-Newton method, but in some cases, if Gauss-Newton has convergence issues, using Levenberg-Marquart may aid the convergence. Practically, Leveberg-Marquart combines Gauss-Newton and gradient descent.

Now using the corollaries from Eqs. (3.5) and (3.6), we can write the iteration in Eq. (6.4) in the standard form

$$\mathbf{w}_j^{i+1} = \mathbf{w}_j^i - \frac{\gamma^i}{1+\lambda^i} \Big\{ \mathbf{w}_j^i - \big(\mathbf{G}_j^i\mathbf{A}\big)^\mathrm{T} \Big(\big(\mathbf{G}_j^i\mathbf{A}\big)\big(\mathbf{G}_j^i\mathbf{A}\big)^\mathrm{T} + (1+\lambda^i)\overline{\mathbf{C}}_{dd} \Big)^{-1} \\ \times \Big(\big(\mathbf{G}_j^i\mathbf{A}\big)\mathbf{w}_j^i + (1+\lambda^i)\big(\mathbf{d}_j - \mathbf{g}\big(\mathbf{z}_j^\mathrm{f} + \mathbf{A}\mathbf{w}_j^i\big)\big)\Big)\Big\}. \tag{6.5}$$

Next, we will replace the tangent-linear model with ensemble covariances.

6.2 Subspace EnRML Algorithm

By introducing the ensemble representation for the covariances in the linear regression Eq. (4.1), we obtain

$$\mathbf{G}_j^i \triangleq \overline{\mathbf{G}}^i = \overline{\mathbf{C}}_{yz}^i \overline{\mathbf{C}}_{zz}^{i\,\dagger} = \mathbf{Y}^i \mathbf{A}^{i\,\dagger}, \tag{6.6}$$

where $\mathbf{Y}^i$ is defined from Eq. (5.9) and evaluated at iteration i, i.e.,

$$\mathbf{Y}^i = \mathbf{g}\big(\mathbf{Z}^i\big)\mathbf{\Pi}. \tag{6.7}$$

The tricky term in Eq. (6.5), which corresponds to the one mentioned in relation to Eq. (4.5), is the product $\mathbf{G}_j^i\mathbf{A}$. Evensen et al. (2019) showed that we can write

$$\mathbf{S}^i = \overline{\mathbf{G}}^i \mathbf{A} = \mathbf{Y}^i \mathbf{A}^{i\,\dagger} \mathbf{A} \tag{6.8}$$

$$= \mathbf{Y}^i \mathbf{A}^{i\,\dagger} \mathbf{A}^i \mathbf{\Omega}^{i\,-1} \tag{6.9}$$

$$= \mathbf{Y}^i \mathbf{\Omega}^{i\,-1} \quad \text{if } n \geq N-1 \text{or if } \mathbf{g} \text{ is linear.} \tag{6.10}$$

Algorithm 1 Subspace EnRML algorithm

Input: $\mathbf{Z} \in \mathcal{R}^{n\times N}$ ▷ Prior state-vector ensemble
Input: $\mathbf{D} \in \mathcal{R}^{m\times N}$ ▷ Perturbed measurements
$\mathbf{W}^{(0)} = 0$ ▷ $\mathbf{W} \in \mathcal{R}^{N\times N}$
$\boldsymbol{\Pi} = \left(\mathbf{I}_N - \frac{1}{N}\mathbf{1}\mathbf{1}^{\mathrm{T}}\right)/\sqrt{N-1}$ ▷ $\boldsymbol{\Pi} \in \mathcal{R}^{N\times N}$
$\mathbf{E} = \mathbf{D}\boldsymbol{\Pi}$ ▷ $\mathbf{E} \in \mathcal{R}^{m\times N}$
$i = 0$
repeat
  $\mathbf{Y}^i = \mathbf{g}(\mathbf{Z}^i)\boldsymbol{\Pi}$ ▷ $\mathbf{Y} \in \mathcal{R}^{m\times N}$
  if $(n < N-1)$ $\mathbf{Y}^i = \mathbf{Y}^i\mathbf{A}^{i\dagger}\mathbf{A}^i$
  $\boldsymbol{\Omega}^i = \mathbf{I}_N + \mathbf{W}^i\boldsymbol{\Pi}$ ▷ $\boldsymbol{\Omega} \in \mathcal{R}^{N\times N}$
  $\mathbf{S}^i = \mathbf{Y}^i{\boldsymbol{\Omega}^i}^{-1}$ ▷ $\mathbf{S} \in \mathcal{R}^{m\times N}$
  $\mathbf{W}^{i+1} = \mathbf{W}^i - \gamma\left(\mathbf{W}^i - {\mathbf{S}^i}^{\mathrm{T}}\left(\mathbf{S}^i{\mathbf{S}^i}^{\mathrm{T}} + \mathbf{E}\mathbf{E}^{\mathrm{T}}\right)^{-1}\left(\mathbf{S}^i\mathbf{W}^i + \mathbf{D} - \mathbf{g}(\mathbf{Z}^i)\right)\right)$
  $\mathbf{T}^i = \left(\mathbf{I}_N + \mathbf{W}^{i+1}\big/\sqrt{N-1}\right)$ ▷ $\mathbf{T} \in \mathcal{R}^{N\times N}$
  $\mathbf{Z}^{i+1} = \mathbf{Z}\mathbf{T}^i$
  $i = i + 1$
until convergence

In this expression, we have defined the quadratic matrix

$$\boldsymbol{\Omega}^i = \mathbf{I}_N + \mathbf{W}^i\boldsymbol{\Pi}, \tag{6.11}$$

which relates the ensemble anomalies at iteration i to the initial anomalies $\mathbf{A} = \mathbf{A}^i{\boldsymbol{\Omega}^i}^{-1}$, and where we have defined the ensemble of weights

$$\mathbf{W} = \left(\mathbf{w}_1,\ \mathbf{w}_2,\ \ldots,\ \mathbf{w}_N\right). \tag{6.12}$$

Note that we cannot use Eq. (6.10) when $n < N-1$, i.e., when the state dimension is less than the ensemble size minus one. We, then, need to retain the projection $\mathbf{A}^{i\dagger}\mathbf{A}^i$ and use Eq. (6.9) rather than Eq. (6.10). Evensen et al. (2019) derived the proofs of this result, and we refer to this paper for the details. This result is also complementary to the use of the expression $\mathbf{Y}\mathbf{A}^{\dagger}\mathbf{A}\mathbf{Y}^{\mathrm{T}}$ to represent $\mathbf{C}_{yy}$ in Eq. (5.13) when $n < N-1$, as was derived by Evensen (2019).

We can now write the iteration of Eq. (6.5) in matrix form as

$$\begin{aligned}\mathbf{W}^{i+1} = \mathbf{W}^i - \frac{\gamma^i}{1+\lambda^i}\Big\{\mathbf{W}^i - {\mathbf{S}^i}^{\mathrm{T}}\left(\mathbf{S}^i{\mathbf{S}^i}^{\mathrm{T}} + (1+\lambda^i)\overline{\mathbf{C}}_{dd}\right)^{-1} \\ \times\left(\mathbf{S}^i\mathbf{W}^i + (1+\lambda^i)\left(\mathbf{D} - \mathbf{g}(\mathbf{Z}^i)\right)\right)\Big\}.\end{aligned} \tag{6.13}$$

The update for iteration i is

$$
\begin{aligned}
\mathbf{Z}^i &= \mathbf{Z} + \mathbf{A}\mathbf{W}^i \\
&= \mathbf{Z}\big(\mathbf{I}_N + \boldsymbol{\Pi}\mathbf{W}^i\big) \\
&= \mathbf{Z}\big(\mathbf{I}_N + \mathbf{W}^i/\sqrt{N-1}\big),
\end{aligned}
\tag{6.14}
$$

where we have used $\mathbf{W}^i/\sqrt{N-1} = \boldsymbol{\Pi}\mathbf{W}^i$, which we get from Eq. (6.13) using $\mathbf{S}^{\mathrm{T}}/\sqrt{N-1} = \boldsymbol{\Pi}\mathbf{S}^{\mathrm{T}}$. Thus, we can compute the final update in Eq. (6.14) to a cost of nN^2 operations. The updated ensemble is a linear combination of the prior ensemble members, and the prior ensemble space contains the updated ensemble of solutions.

Algorithm 1 details the Gauss-Newton implementation of the subspace EnRML algorithm. We excluded the Levenberg-Marquart parameter λ as we have never needed to use it in the ensemble subspace implementation. The algorithm takes as inputs the prior ensemble and the perturbed measurements and runs an ensemble of model simulations to evaluate $\mathbf{g}(\mathbf{Z}^i)$. Thus, the algorithm is generic, and we can use it for any model or problem configuration. In Sect. 6.5, we discuss a practical and efficient implementation for inverting the expression $\big(\mathbf{S}^i\mathbf{S}^{i\mathrm{T}} + \overline{\mathbf{C}}_{dd}\big)^{-1}$, where we replace the full measurement error covariance matrix with the ensemble representation, $\mathbf{C}_{dd} \approx \overline{\mathbf{C}}_{dd} = \mathbf{E}\mathbf{E}^{\mathrm{T}}$.

Choosing an optimal value for the step length γ^i is generally not straightforward. Evensen et al. (2019) proposed a scheme with γ^i decreasing from 0.6 to 0.1. For linear problems, an optimal step length of $\gamma^i = 1.0$ gives us the solution in one step. However, for nonlinear problems where we apply an averaged model sensitivity, we must use a conservative value of γ that leads to stable solutions for all the realizations. We typically start with a value $\gamma^i = 0.5$, and if the sum of the cost functions in Eq. (5.19) increases from one iteration to the next, we divide the step length by two and rerun that iteration. Note that we only need to store the $\mathbf{W}$ and the predicted measurements $\boldsymbol{\Upsilon}$ from the previous iteration and then recompute the updated $\mathbf{W}^i$ from Eq. (6.13). A critical learning is that one should rerun an iteration with a smaller value of γ^i if the mean cost function value increases over an iteration.

6.3 Ensemble Smoother

Interestingly, it is possible to compute the EnKF or ES solution from the first subspace EnRML iteration. Using the ensemble matrices from the previous chapter, we can rewrite the ES update Eq. (5.13), as

$$
\begin{aligned}
\mathbf{Z}^{\mathrm{a}} &= \mathbf{Z}^{\mathrm{f}} + \mathbf{A}\mathbf{S}^{\mathrm{T}}\big(\mathbf{S}\mathbf{S}^{\mathrm{T}} + \mathbf{E}\mathbf{E}^{\mathrm{T}}\big)^{-1}\big(\mathbf{D} - \mathbf{g}(\mathbf{Z}^{\mathrm{f}})\big) \\
&= \mathbf{Z}^{\mathrm{f}} + \mathbf{A}\mathbf{W},
\end{aligned}
\tag{6.15}
$$

with

$$
\mathbf{W} = \mathbf{S}^{\mathrm{T}}\big(\mathbf{S}\mathbf{S}^{\mathrm{T}} + \mathbf{E}\mathbf{E}^{\mathrm{T}}\big)^{-1}\big(\mathbf{D} - \mathbf{g}(\mathbf{Z}^{\mathrm{f}})\big). \tag{6.16}
$$

Note that we have redefined $\mathbf{S}$ as

$$\mathbf{S} = \begin{cases} \mathbf{Y} & \text{for } n \geq N - 1 \\ \mathbf{Y}\mathbf{A}^{\dagger}\mathbf{A} & \text{for } n < N - 1. \end{cases} \tag{6.17}$$

If we set the step length $\gamma = 1.0$ and the relaxation parameter $\lambda = 0$ in the Levenberg-Marquart iteration of (6.13), the first step becomes just the update Eq. (6.15), as the prior value $\mathbf{W}^{(0)} = 0$. Hence, we can use Algorithm 1 to compute both the EnRML and the ES solutions.

6.4 Ensemble Smoother with Multiple Data Assimilation

The Ensemble Smoother with Multiple Data Assimilation (ESMDA) method proposed by Emerick and Reynolds (2013) uses ES updates for each recursive update step in the MDA algorithm from Sect. 3.6. Since the ES update is an ensemble that approximately samples the posterior distribution, we can use the ES-updated ensemble as the prior for the next update step. Furthermore, since we can use Algorithm 1 to compute the ES solution, we can also recursively compute the solution for the ESMDA method.

We use the algorithm as we do for the ES solution, but we repeat the procedure n_α times. As the steps are independent, we must resample the perturbed measurements from $\mathcal{N}(\mathbf{d}, \alpha_i \mathbf{C}_{dd})$ for each recursive call to the algorithm. Thus, each recursive step uses an effective measurement error variance increased by a factor α_i, and we compute a sequence of small update steps that gradually introduce the information from the measurements. Hence, ESMDA computes many short linear steps instead of one long one, reducing the impact of nonlinearity when using the linear ES update equation. Note that ESMDA with one step corresponds to the ES estimate.

There is a need for a better theoretical understanding and basis for choosing the optimal sequence of weights. Using uniform weights, i.e., setting all $\alpha_i = n_\alpha$ which satisfy Eq. (3.17) is standard. However, some works have provided evidence that a geometrically decreasing set of weights gives superior results (Rafiee and Reynolds 2017; Evensen 2018; Emerick 2019).

ESMDA has gained popularity due to its ease of implementation and successful use in many applications. When considering the convergence of ESMDA, we mean the number of update steps needed before a further decrease in step length does not change the final solution. The required number of steps depends on the nonlinearity of the model. The typical number of steps in practical applications ranges from 4 to 16, depending on the model's nonlinearity. We note that in ESMDA, as the number of steps increases, the magnitude of the measurement perturbations also increases. Thus, one can imagine cases where the perturbed measurements take unphysical values, causing the algorithm to break down. Emerick (2018) resolved this issue using a square-root formulation for the update calculation, although the paper's objective was to reduce sampling errors.

Chapter 9 will provide simple scalar examples illustrating some properties of the ESMDA method, and ESMDA is also the method used in the reservoir cases in Chap. 13.

6.5 Ensemble Subspace Inversion

In Algorithm 1, we have represented the measurement error covariance matrix by an ensemble of measurement perturbations, $\mathbf{E}$. In this case, we compute the inversion using an ensemble subspace scheme proposed by Evensen (2004). See also the more recent discussions by Evensen (2009, 2021) and Evensen et al. (2019). The scheme projects the measurement error perturbations onto the ensemble subspace and computes the pseudo inverse using the following factorization

$$\begin{aligned}
&\left(\mathbf{S}\mathbf{S}^{\mathrm{T}} + \mathbf{E}\mathbf{E}^{\mathrm{T}}\right) && (6.18)\\
&\quad\approx \mathbf{S}\mathbf{S}^{\mathrm{T}} + (\mathbf{S}\mathbf{S}^{\dagger})\mathbf{E}\mathbf{E}^{\mathrm{T}}(\mathbf{S}\mathbf{S}^{\dagger})^{\mathrm{T}} && (6.19)\\
&\quad= \mathbf{U}\boldsymbol{\Sigma}\left(\mathbf{I}_N + \boldsymbol{\Sigma}^{\dagger}\mathbf{U}^{\mathrm{T}}\mathbf{E}\mathbf{E}^{\mathrm{T}}\mathbf{U}(\boldsymbol{\Sigma}^{\dagger})^{\mathrm{T}}\right)\boldsymbol{\Sigma}^{\mathrm{T}}\mathbf{U}^{\mathrm{T}} && (6.20)\\
&\quad= \mathbf{U}\boldsymbol{\Sigma}\left(\mathbf{I}_N + \mathbf{Q}\boldsymbol{\Lambda}\mathbf{Q}^{\mathrm{T}}\right)\boldsymbol{\Sigma}^{\mathrm{T}}\mathbf{U}^{\mathrm{T}} && (6.21)\\
&\quad= \mathbf{U}\boldsymbol{\Sigma}\mathbf{Q}(\mathbf{I}_N + \boldsymbol{\Lambda})\mathbf{Q}^{\mathrm{T}}\boldsymbol{\Sigma}^{\mathrm{T}}\mathbf{U}^{\mathrm{T}}, && (6.22)
\end{aligned}$$

where we define the singular-value decomposition

$$\mathbf{S} = \mathbf{U}\boldsymbol{\Sigma}\mathbf{V}^{\mathrm{T}}, \tag{6.23}$$

and the identity matrix $\mathbf{I}_N \in \mathcal{R}^{N\times N}$. We must specify a truncation value in the singular-value decomposition in Eq. (6.23) to avoid dividing by zero when computing the pseudo inverse $\boldsymbol{\Sigma}^{\dagger}$. The truncation value is not critical, and a typical truncation accounts for around 99% of the variance represented by the largest singular values.

The eigenvalue decomposition in Eq. (6.21) is of the matrix product in (6.20). Note that this eigenvalue decomposition is most efficiently computed by a singular-value decomposition of the product $\boldsymbol{\Sigma}^{\dagger}\mathbf{U}^{\mathrm{T}}\mathbf{E}$. The left singular vectors will then equal the eigenvectors in $\mathbf{Q}$, and the squares of the singular values will equal the eigenvalues in $\boldsymbol{\Lambda}$. Thus, the inversion becomes

$$\begin{aligned}
&\left(\mathbf{S}\mathbf{S}^{\mathrm{T}} + \mathbf{E}\mathbf{E}^{\mathrm{T}}\right)^{-1}\\
&\quad\approx \left(\mathbf{U}(\boldsymbol{\Sigma}^{\dagger})^{\mathrm{T}}\mathbf{Q}\right)\left(\mathbf{I}_N + \boldsymbol{\Lambda}\right)^{-1}\left(\mathbf{U}(\boldsymbol{\Sigma}^{\dagger})^{\mathrm{T}}\mathbf{Q}\right)^{\mathrm{T}}. && (6.24)\\
&\quad= \mathbf{U}(\boldsymbol{\Sigma}^{\dagger})^{\mathrm{T}}\mathbf{Q}\left(\mathbf{I}_N + \boldsymbol{\Lambda}\right)^{-1}\mathbf{Q}^{\mathrm{T}}\boldsymbol{\Sigma}^{\dagger}\mathbf{U}^{\mathrm{T}}
\end{aligned}$$

The main advantage of this algorithm is that it allows for computing the inverse to a linear cost in the number of measurements, $\mathcal{O}(mN^2)$. The algorithm represents the measurement error covariances using the measurement perturbation matrix $\mathbf{E}$. Simulating measurement perturbations with given statistics is usually easier than constructing a complete error covariance matrix. The disadvantage is that using a finite ensemble to represent the measurement error covariance matrix introduces additional sampling errors. However, (Evensen, 2021) demonstrated that, by using a larger ensemble to represent $\mathbf{E}$ in Eq. (6.18), one could reduce the associated sampling errors to a negligible magnitude and with little additional computational cost since $\mathbf{E}$ only occur in the matrix multiplication $\mathbf{U}^\mathrm{T}\mathbf{E}$ in Eq. (6.20).

6.6 EnKF Analysis with Independent Measurements

Independent measurements are measurements with uncorrelated measurement errors. Most operational ensemble-based assimilation schemes apply an assumption of uncorrelated measurement errors and use a diagonal $\mathbf{C}_{dd} = \mathbf{I}_m$; see, e.g., the reviews on data assimilation in the geosciences (Carrassi et al. 2018), weather prediction (Houtekamer and Zhang 2016), and petroleum applications (Aanonsen et al. 2009). Data assimilation practitioners employ this assumption for two reasons. First, the measurement error covariances are often unknown, so one avoids modeling them by setting them to zero. Additionally, the assumption of a diagonal $\mathbf{C}_{dd}$ somewhat simplifies the update scheme in Eq. (5.12). With $\mathbf{C}_{dd} = \mathbf{I}_m$, Eq. (6.15) becomes

$$\mathbf{Z}^\mathrm{a} = \mathbf{Z}^\mathrm{f} + \mathbf{A}\mathbf{S}^\mathrm{T}\left(\mathbf{S}\mathbf{S}^\mathrm{T} + \mathbf{I}_m\right)^{-1}\left(\mathbf{D} - \mathbf{g}(\mathbf{Z})\right), \tag{6.25}$$

which makes it possible to use an efficient algorithm proposed by Hunt et al. (2007) where, by using a Woodbury identity, the EnKF-update becomes

$$\mathbf{Z}^\mathrm{a} = \mathbf{Z}^\mathrm{f} + \mathbf{A}\left(\mathbf{S}^\mathrm{T}\mathbf{S} + \mathbf{I}_N\right)^{-1}\mathbf{S}^\mathrm{T}\left(\mathbf{D} - \mathbf{g}(\mathbf{Z})\right). \tag{6.26}$$

This modification reduces the size of the matrix inversion from $m \times m$ in Eq. (6.25) to $N \times N$ in Eq. (6.26). See also the discussion on this implementation in Evensen et al. (2019, Sect. 3.2).

Finally, the subspace inversion from the previous section is equally fast as the "exact" inversion in Eq. (6.26), even for a diagonal measurement error covariance matrix. The authors' experience is that the exact inversion in Eq. (6.26) is less stable than the subspace inversion from the previous section. Evensen et al. (2024), therefore, used the subspace inversion for all experiments.

6.7 Impact of Measurement Error Correlations

Evensen (2021) discussed the impact of conditioning on measurements with correlated errors, and we will now revisit and extend an example from this paper. Figure 6.1 shows the updates from three EnKF experiments, all using 2000 model realizations to minimize sampling errors. The upper plots represent a case where we

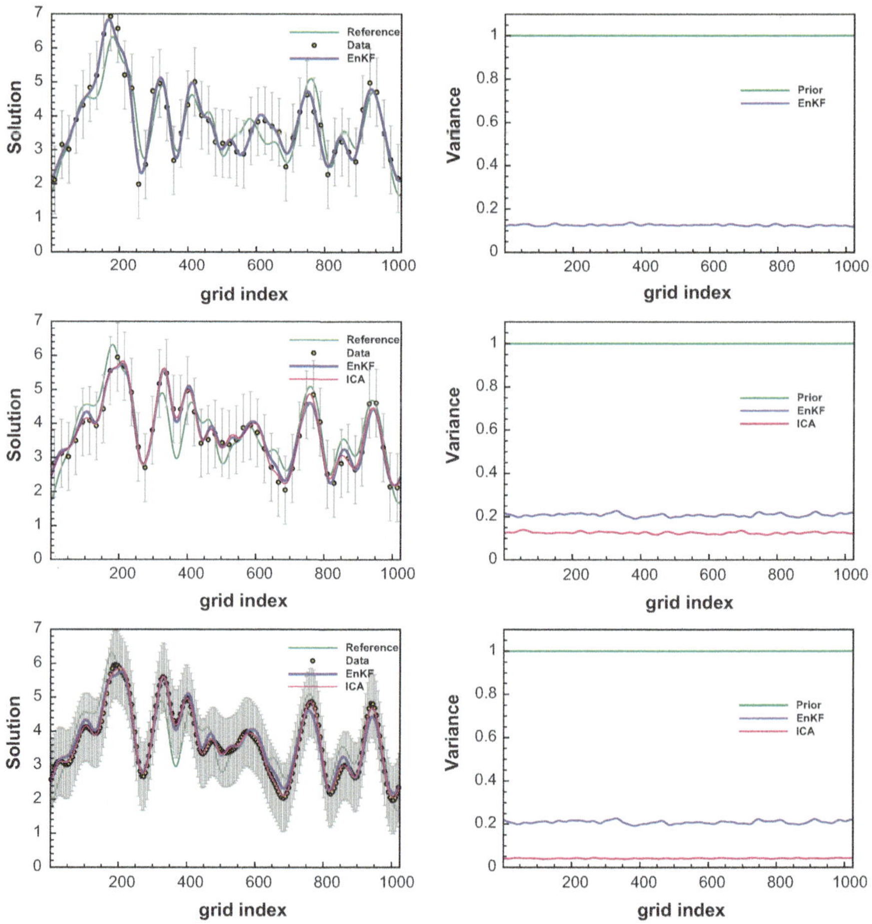

Fig. 6.1 Impact of measurement dependency on analysis update. The upper row presents the results for a case with uncorrelated measurement errors, while the second row gives the results when using measurements with correlated errors. The third row is similar to the second but uses four times as many observations. The left plots show the results for the posterior ensemble means, while the panels to the right provide the associated error variance estimates. The line labels EnKF and ICA denote the standard EnKF update and an inconsistent update, as is explained in the text. The measurement error bars indicate the plus-minus two standard deviations of the measurement errors

condition the prior ensemble on 50 independent measurements with error standard deviations of 0.5. We have plotted the measurements with error bars indicating plus and minus two standard deviations. We obtained an excellent EnKF result that agreed with the measurements given their errors and recovered the reference solution well. The posterior variance varies around 0.125, a significant reduction compared to the prior variance of one.

The experiment shown in the second row is similar to the one in the first row. However, we assume correlated measurement errors with a decorrelation length of 40 (grid indexes). From the EnKF posterior variance shown in the right plot, we notice that introducing measurement error correlations reduces the EnKF-update's strength, and the posterior variance now varies around 0.2. The importance of specifying and accounting for any measurement error correlations becomes evident from the experiment displayed in the third row of Fig. 6.1. This experiment differs from the second row by conditioning on 200 dependent measurements instead of 50. In the EnKF-analysis update, we avoid any overfitting of the observations, and the posterior variance is nearly identical to the experiment with 50 dependent measurements. Thus, because of the dependencies between neighboring observations, increasing the measurement density does not lead to the assimilation of more information or a more substantial update.

An interesting result is the ICA case (ICA denotes inconsistent analysis), where we have conditioned on measurements with correlated errors but neglected these correlations when computing the update, i.e., we are using a diagonal measurement error covariance matrix. In this case, we obtain an underestimated posterior variance similar to the example with uncorrelated measurement errors. In the example with 200 correlated measurements, the variance underestimation is even more pronounced, and we see the potential for ensemble collapse when increasing the number of measurements.

The ICA cases represent the typical approach currently used in most ensemble methods applied by history-matching practitioners. Neglecting the measurements' error correlations leads to overfitting and, eventually, so-called ensemble collapse. For example, when conditioning the model on a time series of rate data, one needs to select the sampling frequency of data in time. For example, should we use weekly, monthly, or another data sampling frequency? And does an increased sampling frequency introduce additional information to the conditioning process? From the example above, it is clear that increasing the sampling frequency of dependent data without representing the measurement error correlations is inconsistent. It leads to underestimating the posterior ensemble variance. We will revert to this discussion in Sect. 13.6.

For the practical computation and representation of measurement error correlations, it is simpler to simulate a time series of correlated errors and store these in the measurement error perturbation matrix $\mathbf{E}$ used to define the ensemble measurement error covariance in Eq. (5.7). Also, in these examples, we used $10N$ realizations of the measurement perturbations to reduce the sampling errors.

Chapter 7
Correlation-Based Localization

Abstract This chapter will introduce an adaptive correlation-based localization method used in the experiments in Part III. The technique is a straightforward implementation of a local analysis scheme where we update the parameters, or rows in the ensemble state-matrix, sequentially one by one using only the significantly correlated measurements in the update. Such adaptive localization has advantages over distance-based localization, particularly when conditioning the model on non-local measurements and when the localization distance is unknown. Another benefit is that an adaptive correlation-based localization does not refer to the spatial location of the updated variables or measurements. Thus, we only need the ensemble matrices of the state and the predicted measurements to determine the localization. We will present a theoretical foundation for selecting the truncation value as a function of the ensemble size and discuss some possible remedies for further incremental improvement of the method.

7.1 Introduction

Localization is a standard technique for reducing the degradation of the quality of updates caused by sampling errors in ensemble-based data assimilation methods. When optimally applied to large-scale problems with small ensemble sizes, localization increases the solution space's degrees of freedom, allowing for a better match to the data used in the conditioning. Furthermore, localization reduces the spurious model updates caused by sampling errors and, thereby, the tendency for the ensemble to "collapse" when we condition on a large number of independent measurements. We refer to Evensen et al. (2022, Chap. 10) for an overview of localization methods and Chen and Oliver (2017) for localization when history-matching reservoir models.

So-called *distance-based localization* is the most commonly used localization method in meteorology and oceanography. The dynamical system's physical scales determine the truncation distance in *distance-based localization*. The dominant scale in the ocean and atmosphere is the mesoscale, which is associated with the Rossby radius of deformation. The procedure for selecting a localization radius, typically a

G. Evensen et al., *Ensemble History Matching*,
https://doi.org/10.1007/978-3-031-99155-4_7

few Rossby radii, must ensure that measurements of, e.g., temperature, sea-surface height, and pressure only influence the model's state variables in a region close to the measurements. For these problems, *distance-based localization* gives results almost equivalent to the infinite ensemble size limit for as long as the measurements are local and we have a well-defined decorrelation length. The first publications on distance-based localization seem to Hamill (2001), Houtekamer and Mitchell (2001), Bishop et al. (2001), Whitaker and Hamill (2002), Anderson (2003), who introduced *covariance localization,* where a tapering function damps long-range spurious correlations. Related to *covariance localization*, we have *Kalman gain localization* where we apply the tapering directly to the Kalman gain (Chen and Oliver 2017).

The truncation distance must be clarified, especially when the observations are non-local or we use elaborate prior reservoir models. In reservoir engineering, we update the reservoir's porosity and permeability fields. Here, the depositional environment that created the geological parameters also determines their decorrelation scales. These correlations persist even when the subsurface undergoes processes like faulting and bending. Thus, a fault separating two reservoir segments does not prevent a well-measurement from one side of the faultline to update the porosity on the other side.

Additionally, while the measurements of atmospheric and oceanic variables are predominantly "local," the production rates from wells are generally "non-local," since the production from one well depends on the reservoir properties within an extended area, e.g., between an injector and the producer. The pressure transients driving the reservoir flow can have a very long range. Thus, the distance-based localization procedure is challenging when history-matching petroleum reservoir models. This conclusion also applies to segment-based localization, where we only use wells within a reservoir segment to update parameters within this segment.

In such complex cases, *adaptive correlation-based localization methods* can help prevent ensemble collapse and reduce data mismatch. Because the adaptive techniques rely on the ability to separate spurious correlations from actual (large ensemble size) correlations, their performance is generally poorer than distance-based localization when distance-based is appropriate.

We will use a version of the so-called *local analysis* introduced by Haugen et al. (2002), Evensen (2003). While Haugen et al. (2002) updated all variables on a small cube of grid points using only nearby located measurements, we will in the following section introduce an *adaptive correlation-based localization* in combination with a local analysis scheme. From the previous chapters, we also note that we never compute the state covariance matrix or the Kalman gain. Instead, we calculate the analysis as a linear combination of the ensemble realizations.

7.2 Adaptive Correlation-Based Localization

Adaptive correlation-based localization is a newer, less researched alternative to distance-based localization. It uses the estimated correlations between "predicted measurements" and the state variables to determine if a measurement shall influence a state variable's update. In an update, we only retain the observations with a correlation above a certain threshold.

The adaptive approach removes the dependence on the physical coordinates and locations for state variables and wells. Several recent publications have discussed applications of adaptive correlation-based localization (Evensen 2009, Chap. 15.5) and Neto et al. (2021), Luo and Xia (2022), Luo and Bhakta (2020), Bishop and Hodyss (2007), Bishop and Hodyss (2009a), Bishop and Hodyss (2009b), Anderson (2007), Anderson (2009), Anderson (2012), Anderson (2016), Zhang and Oliver (2010), Vossepoel et al. (2025).

An issue with adaptive localization is that removing all measurements with an estimated correlation less than the truncation value will also ignore information from measurements with natural but low correlations. However, we would have similar problems when using a distance-based truncation. With correlation-based localization, we will include some measurements with spurious high correlation values. Similarly, with distance-based localization, we will retain some measurements with insignificant information located within the truncation distance.

In the work by Neto et al. (2021), the authors updated all parameters in a block of grid points. They took this approach for computational reasons, as there will be less number of local updates. On the other hand, this algorithm also introduces discontinuities between the local grid blocks since the observations used to update the different grid blocs may vary significantly. The authors resolved this problem by adding tapering, or equivalently measurement-variance inflation, of measurements with correlations near the truncation cut-off value. In this book's examples, we have not divided the model grid into local grid blocks but instead updated each parameter individually. The problem with discontinuities in the updated fields then becomes much smaller.

The primary benefit of adaptive localization is its ease of use. We can choose to apply it or not, without any user configuration, just by switching it on or off. Also, adaptive correlation-based localization was demonstrated for sequential data assimilation in a nonlinear coupled multiscale system by Vossepoel et al. (2025). They found that selecting measurements based on the correlation value gave equally good results as the best distance-based localization.

7.3 Implementation Strategy

We can write the global update using methods like ES, ESMDA, and EnRML, as

$$\mathbf{Z}^{\mathrm{a}} = \mathbf{Z}^{\mathrm{f}}\mathbf{T}, \tag{7.1}$$

where the posterior ensemble in $\mathbf{Z}^{\mathrm{a}} \in \Re^{n \times N}$ is a linear combination of the prior ensemble in $\mathbf{Z}^{\mathrm{f}}$. The transition matrix $\mathbf{T} \in \Re^{N \times N}$ determines the linear combination. Here n is the state dimension, N is the number of realizations, and typically $n \gg N$. The matrix multiplication in Eq. (7.1) is the most expensive computation in the global analysis update, requiring nN^2 floating point operations. However, Eq. (7.1) is highly parallelizable as we can compute each row in $\mathbf{Z}^{\mathrm{a}}$ independently of the others.

The *local analysis* exploits the parallelism of the computation in Eq. (7.1) and we can write it as the n local updates for $l = 1 \dots n$ computed from

$$\mathbf{Z}_l^{\mathrm{a}} = \mathbf{Z}_l^{\mathrm{f}} \mathbf{T}_l, \tag{7.2}$$

where the subscript l denotes row l of $\mathbf{Z}$. In the local analysis, we update each row $\mathbf{Z}_l$ using an individual transition matrix $\mathbf{T}_l$, which for each l may differ in the selection of measurements used in its calculation.

The distance and correlation-based localization implementations become very similar when writing the update as Eq. (7.2). The methods only differ in the selection procedure used to compute which observations to use in each local update. We can interpret correlation-based localization as applying another distance measure to define nearby and distant measurements.

The increased computational cost is an issue common for adaptive and distance-based localization. We must compute many inversions in both approaches to obtain the n local $\mathbf{T}_l$s. Still, for each $\mathbf{T}_l$, we will likely use far less number of measurements in the inversions, which comes at the cost of $\mathcal{O}(m_l N^2)$ operations, where m_l is the number of observations used in each local analysis update. Note also that there may be no measurements for many model parameters, and the transition matrix is just the identity matrix.

In distance-based localization, we need to compute the Euclidian distance between each state variable and each measurement to a cost of order $\mathcal{O}(nm)$. It is often possible to significantly reduce this cost by using clever storage schemes for the measurements and combining clusters of nearby grid points into one distance calculation. A significant additional computation in the *correlation-based localization* is evaluating the correlation function between the state variables and the measurements, which fills a matrix $\mathbf{C}_{zy} \in \Re^{n \times m}$. Its computation requires nmN floating point operations to obtain the covariance $\mathbf{C}_{zy} = \mathbf{A}\mathbf{Y}^{\mathrm{T}}$. To get the correlation matrix, we must also normalize by the standard deviations of $\mathbf{A}$ and $\mathbf{Y}$. This computation significantly increases the cost of calculating which measurements to use for each parameter update. On the other hand, each of the local updates will be less costly than the global one as they use fewer measurements in the inversion, and the final update equation is still nN^2 number of operations. Another primary concern is the storage of $\mathbf{C}_{zy} \in \Re^{n \times m}$, which becomes challenging for large models with many measurements. However, we can compute $\mathbf{C}_{zy}$ recursively row by row while computing the local updates.

Another simplification is possible since we expect many adjacent variables to use the same observations in the local update. Thus, a better strategy is to compute the correlations for reasonably sized blocks of rows of $\mathbf{Z}$, corresponding to adjacent variables and grid points, one at a time, and then identify which observations are active when updating each block of variables. In this approach, we would select a dominant variable in the grid-bloock's center to evaluate the correlations with the measurements. By dividing the state ensemble matrix $\mathbf{Z}$ into subblocks, we may reduce the computational cost by orders of magnitude.

We can significantly reduce the number of inversions by reusing the transition matrices $\mathbf{T}_l$ for all state variables updated with the same set of measurements. We also note that while ES and ESMDA initialize $\mathbf{W} = 0$ in each update, EnRML requires the storage of n matrices $\mathbf{W}_l$ of size $N \times N$. Thus, using ES and ESMDA with correlation-based localization is significantly simpler than using EnRML. In EnRML, we also advise using the initially selected active measurements for all the subsequent iterations. Note again that Neto et al. (2021) used correlation-based localization with EnRML and updated blocks of neighboring state variables to reduce the number of local updates.

7.4 Selecting a Correlation Truncation Value

It is possible to derive a theoretical value for the sample-correlation standard deviation using the Fisher transform (see Eq. 9, Flowerdew 2015). It says that the variable r_{lj} from

$$r_{lj} = \frac{1}{2} \ln \frac{1+\rho_{lj}}{1-\rho_{lj}} = \operatorname{arctanh}(\rho_{lj}), \tag{7.3}$$

evaluated from the correlation ρ_{lj} between the measurement number j and the state variable number l, will be approximately normally distributed with a standard deviation of $1/\sqrt{N-3}$. We need the sample standard deviation of the correlation ρ for adaptive correlation-based localization. However, from Fig. 7.2, we see that the standard deviation estimates of ρ and r are nearly identical for the low correlations we wish to truncate. The Fisher transformation corrects for the skewness in the sample standard deviation's distribution close to a correlation of one.

In the case of physically uncorrelated measurements and state variables, the finite ensemble size will lead to spurious correlations with a normal distribution $\mathcal{N}(0, 1/(N-3))$, i.e., with zero mean and a standard deviation of $1/\sqrt{N-3}$. To remove 99.7% of these spurious correlations, we can truncate all measurements with predicted correlations to a state variable of less than three standard deviations (see Fig. 7.1). The choice of three standard deviations makes sense since we wish to remove "all" spuriously correlated measurements, so, for now, we set the truncation to

$$\rho_{\text{trunc}} = 3/\sqrt{N}. \tag{7.4}$$

Figure 7.3 illustrates the truncation of measurements with different levels of true physical correlation. We use an ensemble size of $N = 100$ and a truncation defined by Eq. (7.4) and assume its distribution is approximately Gaussian, which is a

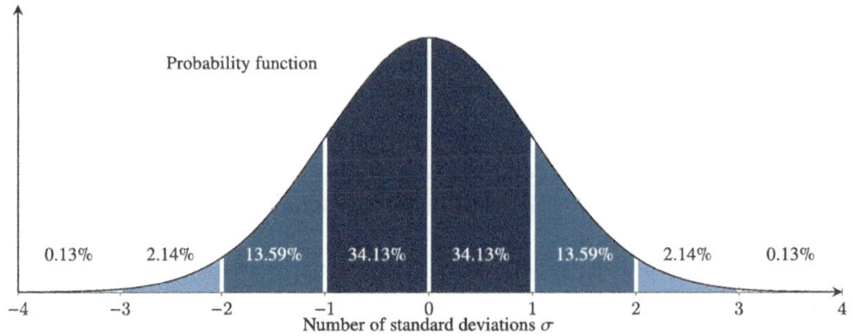

Fig. 7.1 Percentages of samples falling in different parts of the Gaussian pdf

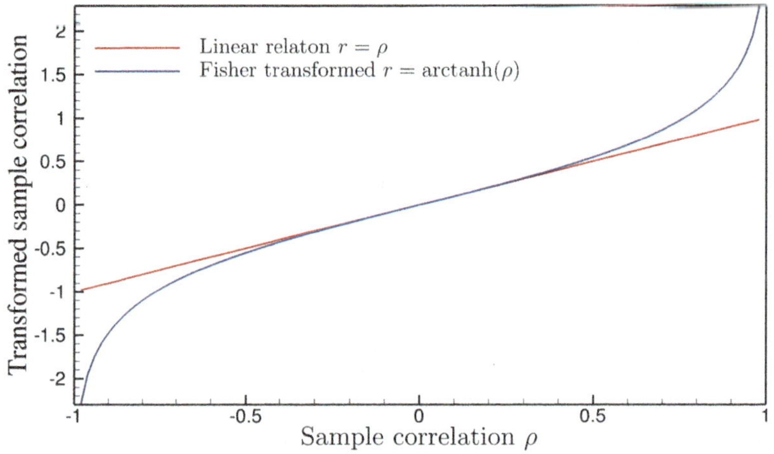

Fig. 7.2 Fisher transformation from Eq. (7.3). Note that the transformation is negligible for correlations in the interval $[-0.6 : 0.6]$ where we apply the correlation-based localization. Thus, the Gaussian assumption on the distributions of sample correlations in Fig. 7.3 is valid

valid approximation in the range of correlation values we consider, see Fig. 7.2 and (Flowerdew 2015, Fig. 7.1).

In the upper plot of Fig. 7.3, we assume zero physical correlation between a measurement and the state variable, as would be true for many of the model's variables. When using a truncation value defined by Eq. (7.4), we will remove an uncorrelated measurement in 99.7% of the cases when truncating based on the sample correlation. The third plot of Fig. 7.3 illustrates that when the measurement has a low physical correlation of 0.2, we remove the measurement in 84.1% of the cases. An alternative interpretation is that with many measurements with a correlation value of 0.2, we will retain 15.9% of them. When the correlation value is 0.3, we will keep half the measurements.

By applying this truncation scheme, we will remove some measurements with a higher physical correlation value located in the range $(\rho_{\text{trunc}} : 2\rho_{\text{trunc}})$, as we truncate

Chapter 8
Non-gaussian and Categorical Variables

Abstract Ensemble methods work with continuous Gaussian distributed parameters and cannot directly handle non-Gaussian or discrete ones. However, it is straightforward to represent non-Gaussian parameters with an underlying Gaussian representation followed by a transformation into the non-Gaussian parameters. We can then use the ensemble methods without violating the Gaussian assumption, but at the disadvantage of introducing additional nonlinearity through the nonlinear transformation. This chapter will discuss the formulation of the Truncated Pluri-Gaussian Simulation (TPS) and Adaptive Pluri-Gaussian Simulation (APS) methods for modeling categorical facies variables.

8.1 Introduction

A fundamental assumption in the derivation of the ensemble methods is the Gaussian Approximation 4 in Chap. 2. This approximation assumes that all state variables are the Gaussian distributed, which allows us to go from Bayes' theorem to a quadratic cost function formulation. However, several critical parameters are not well-modeled as Gaussian in reservoir history matching. Examples include log-normal distributed permeability fields and fault transmissibility distributed uniformly or log-uniformly over the interval from zero to one. The standard approach for handling these non-Gaussian variables in the ensemble history matching is to represent them using underlying Gaussian distributions as the estimated state variables. We then derive the model parameters using different transformations from the Gaussian to their actual distribution. This approach ensures that we do not violate the Gaussian Approximation 4 but with the cost of adding a nonlinear function to the chain of model steps from the Gaussian parameter representation to the predicted measurements. Still, suppose the total nonlinearity is weak, and a "moderate" change of the input Gaussian field leads to a monotonic response in the predicted measurements. In that case, using the iterative ensemble assimilation methods leads to consistent and improved Gaussian parameter representation updates. This approach for handling non-Gaussian parameters in the history matching has been typical since the first ensemble history-matching papers for reservoir simulation models (Evensen et al. 2007; Haugen et al. 2008 e.g.,).

G. Evensen et al., *Ensemble History Matching*,
https://doi.org/10.1007/978-3-031-99155-4_8

On the other hand, the facies estimation adds another level of complexity to the history-matching problem. Not only are the facies fields non-Gaussian distributed, but they are also discrete or categorical variables that change abruptly, e.g., from sand to shale, from one grid point to the next. Fortunately, there is a well-developed framework, i.e., the Truncated Pluri-Gaussian Simulation (TPS) method, for representing different facies configurations using underlying Gaussian fields in combination with a truncation scheme. The following section discusses the standard truncated Gaussian and pluri-Gaussian simulation methods in some detail. After that, Sect. 8.3 introduces probability fields in the TPS method, leading to an even more valuable and flexible facies representation suitable for history matching: the Adaptive Pluri-Gaussian Simulation (APS) method used on the Troll field in Chap. 14.

8.2 Truncated Pluri-Gaussian Simulation

The Truncated Pluri-Gaussian Simulation (TPS) method generates facies realizations using a truncation map to project Gaussian random fields into discrete fields. The truncation map can be user-defined, based on prior geological knowledge (Liu and Oliver 2005), or inferred from the geological model (Sebacher et al. 2013). TPS is a generalization of the Gaussian truncation method (Matheron et al. 1987), where one truncates a single Gaussian field using predefined thresholds on the real axis. Figure 8.1 illustrates the Gaussian truncation method where one simulates a Gaussian field, then applies a simple truncation rule and obtains a facies realization. In Fig. 8.1, the grid cells with Gaussian random field (GRF) value above zero get assigned the sand facies type, and correspondingly, the grid cells with Gaussian value below zero will have a shale facies type. We can shift the zero truncation value to any other number to change the fractions of shale and sand in the reservoir. However, the current scheme is unable to represent more complex facies models.

TPS applies truncation rules that combine several Gaussian random fields into one facies realization, i.e., using more than one Gaussian field and a higher dimensional truncation method. First introduced by Galli et al. (1994) and further developed by Le Loc'h et al. (1994) and Le Loc'h and Galli (1997), the TPS method initially truncated two Gaussian random fields using a map defined in two-dimensional real space. The vertical and horizontal lines intersect to divide the space into rectangular subdomains, with each subdomain assigned a specific facies type. The approach leads to different facies distribution realizations by varying the geostatistical properties of the Gaussian fields, reflecting the distinct characteristics of each facies type.

In Fig. 8.2, we have shown how to transform two different Gaussian random fields into facies realizations. To the left, we display the two Gaussian GRF1 and GRF2 fields with different variograms. In the center, we show two different truncation maps and truncation rules; to the right, we plot the corresponding facies realizations.

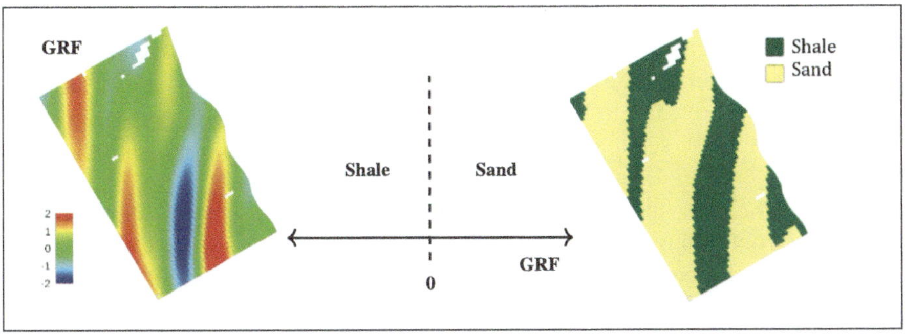

Fig. 8.1 The figure demonstrates how to transform the continuous Gaussian random field (GRF) to the left into the discrete facies map to the right using the truncation rule in the center. Values larger than zero indicate sand, and values below zero indicate shale. We generated the contour plots presented in this chapter using the RMS software from Emerson and AspenTech

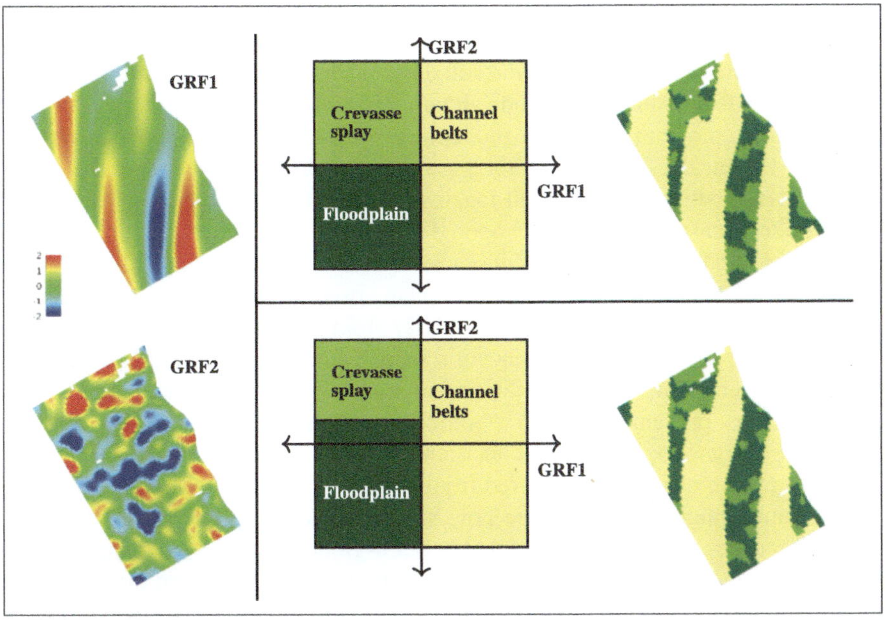

Fig. 8.2 The lower plot illustrates a pluri-Gaussian case where we use two GRFs and a two-dimensional truncation map to create realizations. The example shows how a change in the truncation map will impact the facies realizations and, in this case, the crevasse splay and floodplain distribution

For the upper truncation maps, as long as the GRF1 field has a value above zero, we put channel belt facies in those cells. If the GRF1 is below zero, we apply a similar check of the GRF2. If the value is above zero, we put the crevasse splay facies in the grid cell; if it is below zero, we put the floodplain facies in the grid cell.

We have slightly changed the truncation rule for crevasse splay versus floodplain in the lower truncation map. If GRF1 is above zero, we still put channel belts in the facies realization. However, if it is below zero, we check GRF2; this time, the value must be above 0.3 to have crevasse splay. Compare the two realizations (top and bottom right) to see that the fraction of crevasse splay within the floodplain has decreased.

This method fits the ensemble framework nicely since the facies realizations depend on continuous and Gaussian fields that we can update using the correlations between the fields and the simulated data. Truncated Gaussian methods lead to facies realizations that honor the geological concept as long as a combination of Gaussian fields and truncation rules represents the facies concept. The flexibility of the method is restricted only by the shapes and patterns we can represent in the Gaussian fields. Moreover, the method fits a multiple realization framework because we simulate the Gaussian fields according to a statistical prior.

One of the critical challenges in the TPS method is conditioning facies fields to "hard" data, such as facies observations collected at exploration wells (e.g., Lantuéjoul 2002; Armstrong et al. 2011). Emery (2007, 2008) studied practical considerations for implementing TPS with hard and soft data and developed computer programs for conditioning facies simulations and validating algorithms. Due to its straightforward implementation within the inverse modeling processes, TPS was applied in synthetic fields by Le Ravalec-Dupin et al. (2004) and in real fields by Deraisme and Farrow (2005). It is possible to enforce well-log observations in the prior ensemble iteratively, but the constraints may fail to be honored during production data assimilation. One approach to honoring log data of the facies types is to use the Ensemble Smoother to update the ensemble of Gaussian fields using pseudo-observations corresponding to the facies type, i.e., define (GRF1, GRF2) observations with uncertainty located at appropriate locations in the truncation map. The updated ensemble of Gaussian fields will then change to honor the assimilated data and reduce uncertainty at the data locations, preventing further significant updates at these locations. Thus, the ensemble framework provides a consistent but ad-hoc method for conditioning on direct facies measurements observed from log data.

It is not straightforward to see how we can make the TPS method honor the facies fractions, which must be global or semi-global. An essential aspect of prior information in facies fields involves the potential contacts between different facies types. Xu et al. (2006) tackled this issue using a binary dynamic contact relation matrix within the pluri-Gaussian simulation framework.

Finally, explaining the relation between the variograms in the Gaussian fields and the ad-hoc truncation map is complicated. Le Loc'h and Galli (1997) explored the relationship between the variogram of Gaussian fields and the indicator variogram of facies for the case of stationary proportions. They also introduced a conditional simulation technique for handling non-stationary facies proportions.

Liu and Oliver (2005) introduced the first integration of the TPS method within ensemble-based history matching. Their approach automatically adjusted the facies boundaries when they updated the Gaussian random fields with new measurements. They used the ensemble Kalman filter (EnKF) and randomized maximum likelihood as the history-matching methods, comparing their performances with EnKF, which gave the best results. The facies simulation model employed a truncation map with three lines intersecting in two-dimensional space to create seven regions, each assigned a facies type.

This same truncation map, alongside EnKF, was later used by Agbalaka and Oliver (2008) in a three-dimensional model with three facies types, where any two facies types could be in contact. They also incorporated a distance-based localization scheme. After assimilating production data, they modified the Gaussian field values, resulting in facies fields that sometimes did not match the observed facies types. They applied an additional iterative procedure to correct this mismatch. The facies observation operator used a proxy function that returned zero if the facies matched the observation and one if not.

Zhao et al. (2008) expanded the truncated pluri-Gaussian model introduced by Liu and Oliver (2005) into an iterative EnKF process, integrating both production and seismic data for a three-dimensional reservoir model with three uncorrelated vertical layers. Agbalaka and Oliver (2009) applied EnKF to facies fields exhibiting non-stationary facies proportions while simultaneously updating the petrophysical properties of the facies.

Astrakova and Oliver (2014) applied TPS coupled with an iterative ensemble smoother, incorporating an interior-point method for handling inequality constraints. The interior-point formulation ensured they maintained the correct facies types at well locations.

8.3 Adaptive Pluri-Gaussian Simulation

So far, we have presented the truncated pluri-Gaussian simulation method and learned that the technique simulates realistic facies realizations. This section will introduce the Adaptive Pluri-Gaussian Simulation (APS) method, a facies modeling algorithm based on the TPS method. However, APS introduces an adaptive truncation map that varies at each grid cell according to the prior information about the geology.

Sebacher et al. (2013) introduced a *probabilistic operator for facies observations* and derived the TPS truncation map based on a maximization criterion. They also developed the concept of using *probability fields* to estimate each facies type and with *truncation map parameters incorporated into the state vector* for improved uncertainty quantification. This approach led to the APS method, which represents the categorical nature of facies fields by using underlying Gaussian random fields, like in TPS, together with a *spatially varying truncation map* to define a facies type at each grid cell. Only the underlying Gaussian random fields are random, while we define possibly different truncation maps for each grid cell, which we use for all realizations. We obtain the *spatially varying truncation map* from *facies probability fields,* which we define on the reservoir grid.

A *facies probability field* contains spatial information about the probability of having a specific facies at a particular grid point. We can gather this type of infor-

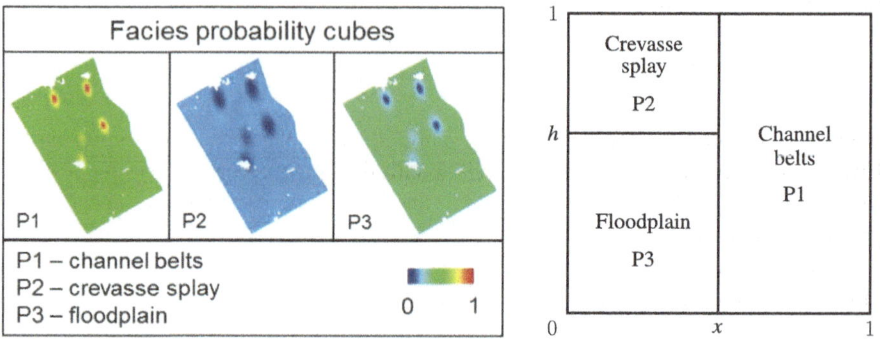

Fig. 8.3 The left panel shows an example of probability cubes for facies distributions conditioned on well-log data. Note how three wells have probability one for channel belts and zero for crevasse splay and floodplain. The right panel is a probability truncation map for a model gridpoint where x and h comes from Eqs. (8.2) and (8.3)

mation from different sources. Well-log data will give us a highly accurate facies log on a fine scale. Inverted seismic can provide a good indication over a large area, although the uncertainty is significantly higher than the well-log data. Outcrops give information on the geological concept. Thus, we start by constructing normalized *facies probability fields* for all facies on the reservoir grid that include all prior geological knowledge of the reservoir, see Fig. 8.3. We assume that the facies probability fields are normalized such that the probability of having a facies in a cell sums to one.

Next, we simulate the ensemble of pairs, y_1, y_2, of *Gaussian random fields* with variogram ranges reflecting the geological concept, similar to the approach used in the TPS method. An essential difference from TPS is that the unit square truncation map lives in the closed $[0, 1] \times [0, 1]$-space (see Fig. 8.3 for an example truncation map), while the Gaussian fields live on the infinite axes. The unit square is similar to the truncation maps used in TPS, but we need to relate the Gaussian random fields to the truncation map, defined with values between zero and one. Thus, we transform each Gaussian random field to a $[0, 1]$-space, using the normal cumulative distribution function for the probability integral transform (Casella and Berger 2024):

$$\alpha(y) = \frac{1}{\sqrt{2\pi}} \int_{-\infty}^{y} e^{-t^2/2} dt, \tag{8.1}$$

see Fig. 8.4. We now have the Gaussian fields and a truncation map in the same space, see Fig. 8.5. By labeling the first and second axes as the α_1-axis and the α_2-axis, respectively, there is a direct link between the GRF fields, y_1 and y_2, and the truncation map. Thus, it is possible to use the TPS strategy, and we have a method from the TPS family using an adaptive truncation map.

Figure 8.3 shows an example of three probability fields conditioned on well-log data that indicate a high probability for channel belts and a low probability for

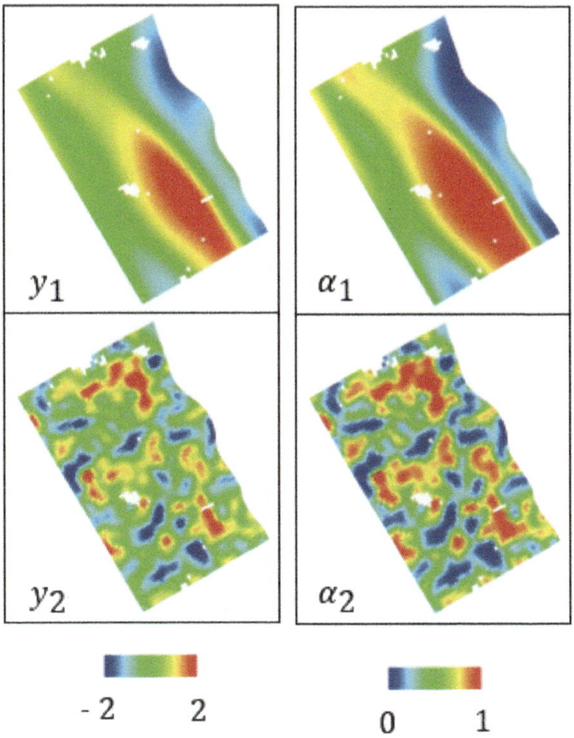

Fig. 8.4 Transform the Gaussian field to a closed [0,1]-domain

crevasse splay and floodplain for three wells. There has to be only one facies type per cell. The unit square's area property and the unit cube's volume property share this property of summing to one. So, for a given cell, it is possible to represent the facies probability fractions as different regions in, say, the unit square, see Fig. 8.3. As long as we place the lines according to the formulas for x and h respectively, i.e.,

$$x = 1 - P1, \tag{8.2}$$

$$h = 1 - \frac{P2}{1 - P1}, \tag{8.3}$$

where P1 and P2 are the probabilities of having channel belts and crevasse splay, respectively, see Fig. 8.3, The area associated with the facies type equals the probability of having that facies in the grid cell. This template truncation map will change for different grid cells according to the facies probability values in each cell, which defines the adaptiveness of the APS method. If a well has a probability of one for one of the facies, the entire truncation map will consist of only one region, the facies observed in the facies log. Thus, while the TGS method honors hard data by updating the Gaussian random fields and their uncertainty, the APS method updates the facies probabilities and uses these to change the truncation map. Still, the outcome is

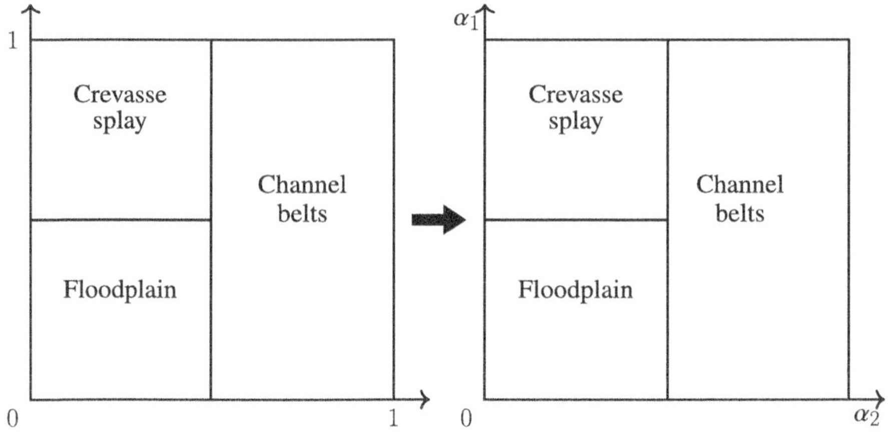

Fig. 8.5 Labeling the truncation map creates a direct link to the "α"-fields

similar, although APS possibly provides greater flexibility and ease in representing more general prior facies models.

Thus, in the example above with three facies types and the truncation map in Fig. 8.5, the procedure for using APS is as follows. For each grid cell (i, j, k), evaluate the probabilities P_1 and P_2 to create the truncation map from Eqs. (8.2) and (8.3). Then, compute α_1 and α_2 from the GRF fields y_1 and y_2, and determine the facies type from the truncation map.

Sebacher et al. (2015), Sebacher et al. (2016) applied the APS method in the context of multi-point geostatistical simulation (MPS) (Caers and Zhang 2004) where random fields, only marginally Gaussian, were truncated using different truncation maps. These random fields were employed to parameterize complex channelized reservoirs generated with MPS and a training image. Unlike standard TPS, these fields were marginally defined by sampling from a standard Gaussian distribution, with truncation maps derived from the probability fields of facies calculated from an ensemble of MPS realizations.

Chapter 9
Nonlinearity Effects

Abstract Nonlinearity is ubiquitous in the relationship between production data and reservoir properties. In most cases, it makes the posterior pdf for inverse problems non-Gaussian, so data assimilation methods that work for linear problems may not be suitable. As discussed in the previous chapters, we can classify the ensemble-based data assimilation methods into two distinct families: the MDA family that repeatedly assimilates the same data many times but with inflated observation error, and the iterative ensembles smoother based on minimization of stochastic objective functions. Both methods provide approximate sampling when the data assimilation problem is nonlinear, but the approximation type differs. In this chapter, we examine the sampling properties for a range of degrees of nonlinearity to better understand the methods' limitations.

9.1 Introduction

Before one can claim to understand a method, it is necessary to understand its limitations. In Chap. 6, we discussed two ensemble-based approaches to assimilating data when the relationships between the model's state vector and the data are nonlinear. The first method is an ensemble-based approximation of the randomized maximum likelihood method (EnRML) that uses optimization for sampling. For nonlinear inverse problems, the distribution of samples obtained using EnRML is only approximate. In the nonlinear case, EnRML violates the Gaussian Approximate 4, and the RML Approximate 6 would require a weighting of the realizations to obtain the correct result (Ba et al. 2022). Since it is rarely feasible to compute the weights in real problems, the distribution of samples from standard, unweighted EnRML is only approximately correct for nonlinear inverse problems. In addition, EnRML applies the linear regression Approximate 7 to represent the individual model sensitivities for each realization with a common averaged one, which may cause problems for the iterative EnRML method with nonlinear models. Consequently, the gradient estimate for an individual model realization can be very poor, and all realizations may converge to a single mode of the posterior distribution.

G. Evensen et al., *Ensemble History Matching*,
https://doi.org/10.1007/978-3-031-99155-4_9

The second method is the ensemble form of multiple data assimilation (ESMDA, see Sect. 6.4). In Sect. 3.6, we showed that the factorization of the likelihood is exact as long as the sum of the inflation factors, $1/\alpha_i$ equals one. For linear data assimilation problems, the ESMDA updates do not change the Monte Carlo approximation of the posterior pdf when using multiple updates with the actual observation set and properly inflated observation errors. However, the ESMDA method becomes an approximation for nonlinear models since it uses the linear ES update equation to compute the distribution of updated state variables from the previous update step. Each ES step will suffer from the approximations applied to derive the EnRML method and the linearization from Approximate 5 used to derive the ES update equation. The method may stall for nonlinear models even when the actual posterior distribution appears to have a single mode.

9.2 A Simple Nonlinear Model

The following single-variable examples illustrate some limitations imposed by the use of ensemble methods for nonlinear data assimilation problems by comparing the evolution and final distribution of samples for three different data assimilation methods: the non-iterative Ensemble Smoother (ES), the iterative Ensemble Randomized Maximum Likelihood (EnRML), and the Ensemble Smoother with Multiple Data Assimilation (ESMDA). We use three nonlinear observation operators, introduced by Chen and Oliver (2013), with varying degrees of nonlinearity, to illustrate the effects of nonlinearity and multi-modality on performance.

$$g_L(x) = 8x \qquad \text{Linear (monotonic)} \tag{9.1}$$

$$g_0(x) = \frac{2}{12}x^3 - \frac{2}{2}x^2 + 8x \qquad \text{Slightly nonlinear (monotonic)} \tag{9.2}$$

$$g_1(x) = \frac{7}{12}x^3 - \frac{7}{2}x^2 + 8x \qquad \text{Moderately nonlinear (monotonic)} \tag{9.3}$$

$$g_2(x) = \frac{20}{12}x^3 - \frac{20}{2}x^2 + 8x \qquad \text{Highly nonlinear (non-monotonic)} \tag{9.4}$$

Figure 9.1 illustrates the properties of the three nonlinear measurement functionals and their implications for the data assimilation problem. From the upper plot, $g_L(x)$, $g_0(x)$, and $g_1(x)$ are monotonic functions of x, while the highly nonlinear function $g_2(x)$ is non-monotonic. All three functions also pass through the points $(0, 0)$ and $(6, 48)$. Non-monotonic measurement functionals pose problems for uncertainty quantification, in general, and for ensemble-based data assimilation methods, in particular, as discussed by Chen and Oliver (2013), Evensen (2018), and (Evensen et al. 2022 Chap. 18).

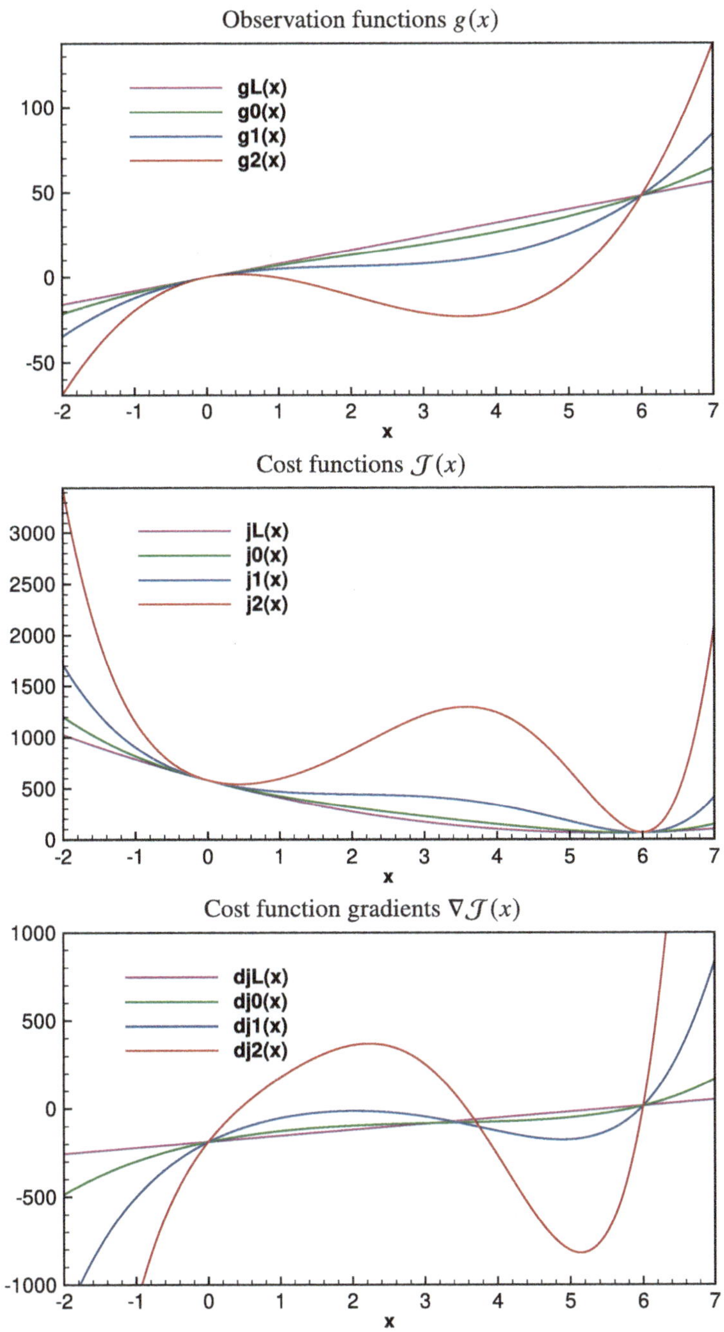

Fig. 9.1 The upper panel displays the observation operators from Eqs. (9.1)–(9.4) used in the numerical examples. The middle panel shows the resulting cost functions from Eq. (9.5), and the bottom panel plots the cost function gradients

For the assimilation experiments below, we assume a prior distribution for the model variable x to be the Gaussian $\mathcal{N}(-2, 1)$ with the mean at -2 and a variance equal to 1. The unknown true value of x equals 6, which is far outside the plausible region based on the prior distribution. We observe a nonlinear function of the true variable for each experiment with an unbiased Gaussian measurement error, $d \sim \mathcal{N}(g(6), 4)$, where the observation functionals evaluated at $x = 6$ equals 48 in all three cases and the measurement errors' standard deviation equals 2.

For the data assimilation experiments, the cost function (see Fig. 9.1, middle panel) is

$$\mathcal{J}(x) = \frac{(x+2)^2}{1^2} + \frac{(g(x) - 48)^2}{2^2}, \tag{9.5}$$

where the function $g(x)$ is one of $g_L(x)$, $g_0(x)$, $g_1(x)$, and $g_2(x)$, from Eqs. (9.1–9.4). All cost functions have minima in the range $x \in (5.5, 6.0)$, although $\mathcal{J}_2(x)$ has a second local minimum near $x = 0$. Greedy optimization algorithms use the gradient or an approximation of the gradient to locate the minimizers. We note in Fig. 9.1 (bottom) that the magnitude of the gradient of the objective function for $g_2(x)$ varies greatly in the interval between the prior mean and the maximum a posteriori (MAP) point, with regions of near-zero gradient and regions in which the gradient will drive a descent-based algorithm away from the MAP location. The situation for $g_1(x)$, while not as bad, has regions of nearly zero gradient far from the MAP.

As $g_1(x)$ is monotonic, there are no local minima in the cost function $\mathcal{J}_1(x)$. However, as the cost function becomes relatively flat close to $x = 2$, its gradient is close to zero, and hence, the cost function's gradient also becomes close to zero around this location. This behavior implies that the gradient methods may converge slowly for particles in the neighborhood of $x = 2$. Note also that with the prior centered around $x = -2$, all the particles must propagate through this area with a low gradient value.

For the highly nonlinear observation function, $g_2(x)$, we obtain a cost function having a local minimum located just above $x = 0$ and the global minimum at $x = 5.99$. We now have a situation where particles must propagate through this local minimum toward the global minimum, which becomes an impossible task for any gradient method. We also note that the high relative value of the cost function at the local minimum compared with the global one yields a posterior probability density function with no visible or significant value around the local minimum.

Finally, we also note that the width of the cost function's "well" around the global minimum determines the posterior uncertainty, and in our case, the posterior uncertainty is the smallest for $g_2(x)$ and then gradually increases for $g_1(x)$ and then $g_0(x)$.

In the following experiments, we evaluate the consequences of applying ES, EnRML, and ESMDA. For ESMDA, we indicate the number of MDA steps used as part of the method name, e.g., ESMDA with four assimilation steps becomes ESMDA_4. For this low-dimensional problem, we can compare results with the actual posterior distribution, which we denote as Bayes in the figure legends. Also, to reduce the effects of Monte Carlo sampling errors, we use a large ensemble size.

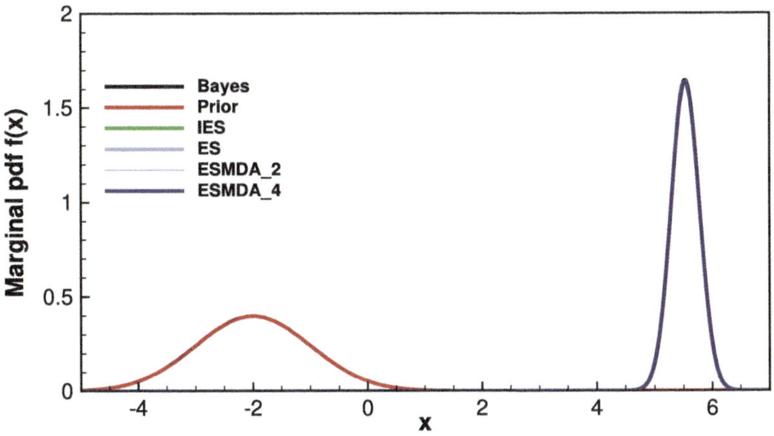

Fig. 9.2 Posterior estimate using the linear function $g_L(x)$

All methods start by drawing samples from the prior and measurement distributions except for the ESMDA method, which samples the observations at each assimilation step from $\mathcal{N}(g(6), 4\alpha)$. Although, for practical purposes, the prior pdf for x is inconsistent with the data in the numerical examples, the two ensemble-based data assimilation methods can generally approximate the posterior distribution for some of the test functions.

9.3 Assimilation of a Linear Observation

For the linear observation operator, $g_L(x)$, all methods, including ES, give results that are indistinguishable from the true Bayes distribution (Fig. 9.2). The EnRML method was relatively inefficient for this problem because it started with a large value $\lambda = 1000$ of the Levenberg-Marquardt damping parameter, which was only allowed to decrease by a factor of 4 at each iteration. The efficiency of the ESMDA method is given by the number of data assimilation steps we specify before initiating the algorithm. The results with any number of steps are identical.

9.4 Assimilation of a Weakly Nonlinear Measurement

For the function g_0, which is only slightly nonlinear in the domain of interest, EnRML converges after about 10 iterations to an excellent approximation of the true posterior pdf (compare the green and black pdfs in Fig. 9.3). The slight variations in the gradient

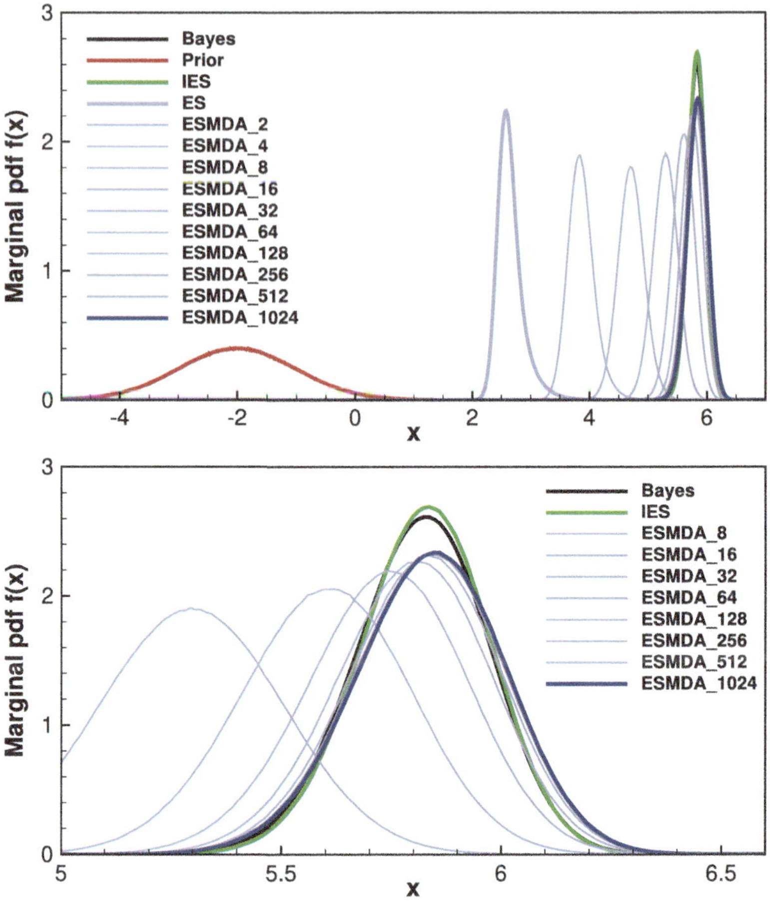

Fig. 9.3 Posterior estimate using the weakly nonlinear function $g_0(x)$

do not appear to cause difficulty in the convergence as shown in the left panels of Fig. 9.4.

The ESMDA algorithm does not perform as well as the EnRML algorithm for this test problem. The distribution of samples from ESMDA with 32 equal assimilation steps was somewhat biased, although probably acceptable in most practical applications with small ensemble sizes. For 64 steps and more, the results from ESMDA change very little (Fig. 9.3). Although the nonlinearity is relatively small in this example, the cost function deviates significantly from quadratic when the measurement error variance is large. This deviation has a noticeable effect on the cost function at intermediate data assimilation steps for ESMDA, causing the particles to advance towards the posterior distribution more slowly than they should (Fig. 9.4, right panels). Because the posterior distribution after an update step in ESMDA becomes the prior for the next update step, update problems at early steps will propagate from

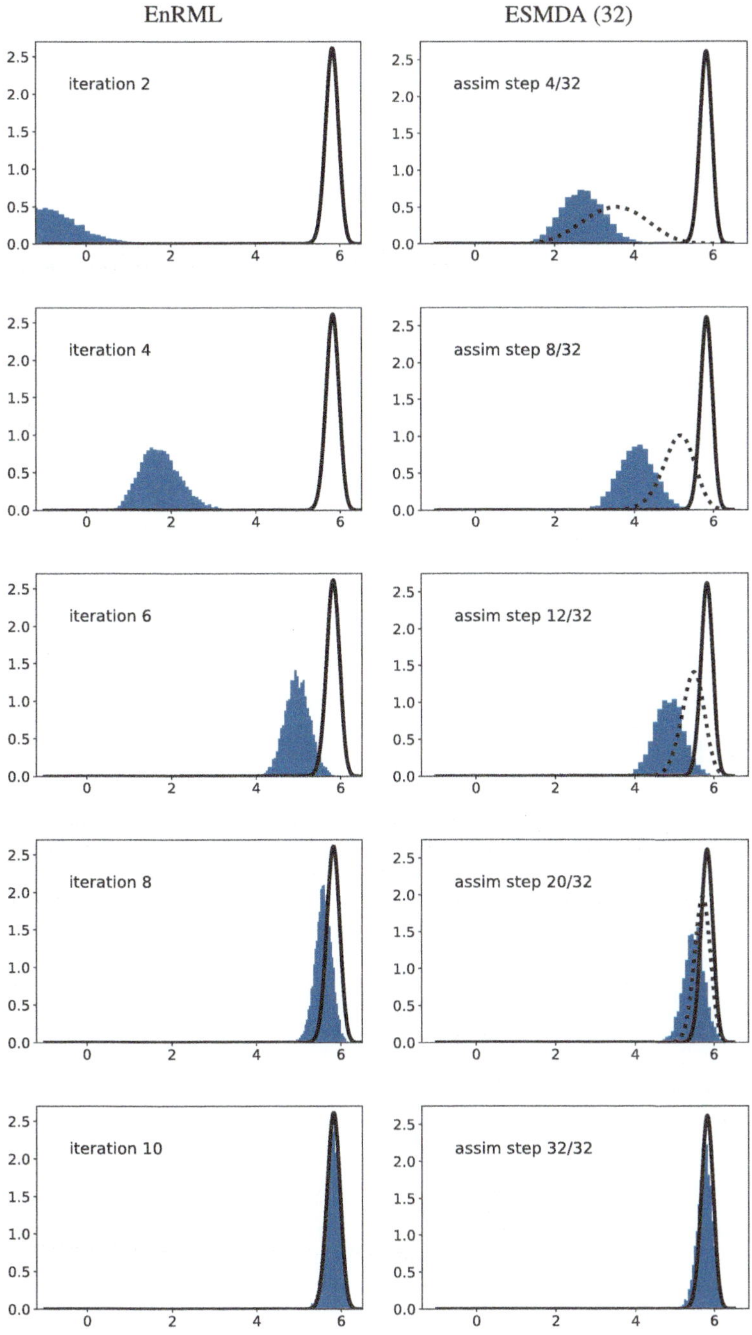

Fig. 9.4 Ensemble-based data assimilation for a slightly nonlinear observation function $g_0(x)$. The dashed line is the true pdf for partially assimilated data. It is the distribution that the histogram should match. At assimilation step 32, it is identical to the target posterior pdf

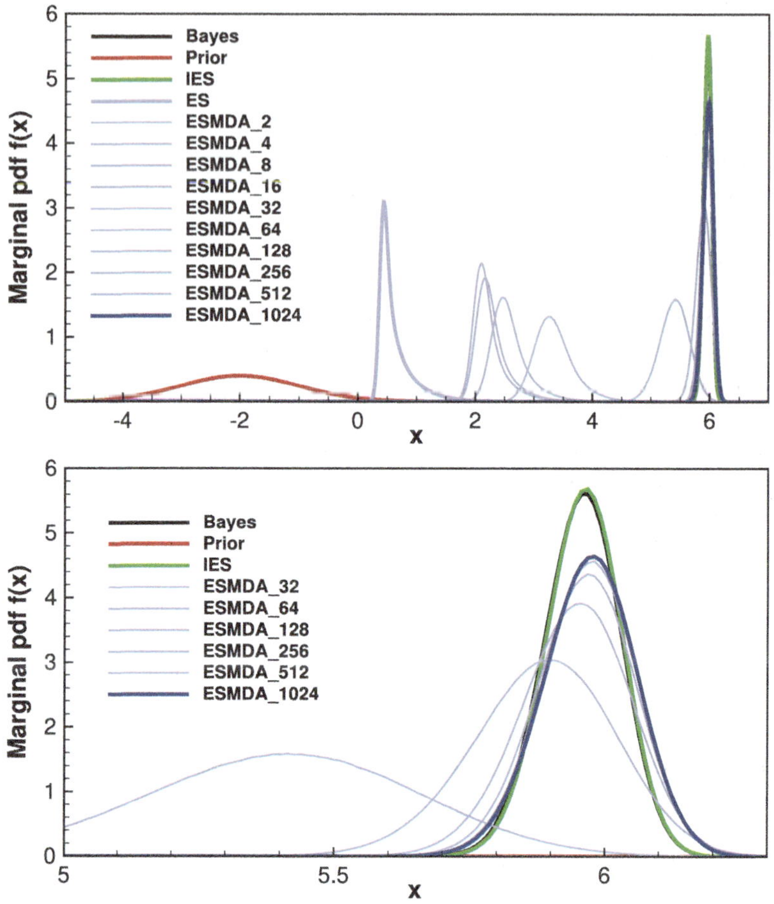

Fig. 9.5 Posterior estimate using the nonlinear function $g_1(x)$. In this case, the EnRML solution (green curve) falls on top of the true posterior pdf

step to step, and the method will not be able to recover at later steps. For this slightly nonlinear example, however, the results are relatively good when using 32 or more data assimilation steps.

9.5 Assimilation of a Moderately Nonlinear Measurement

Like the previous example, the observation function g_1 is monotonic, but in this case, the observation function has a relatively small derivative at $x = 2$, resulting in slower convergence. The nonlinearity is higher than for g_0 due to the more substantial influence of the cubic and quadratic terms in the polynomial. In Fig. 9.6, the EnRML

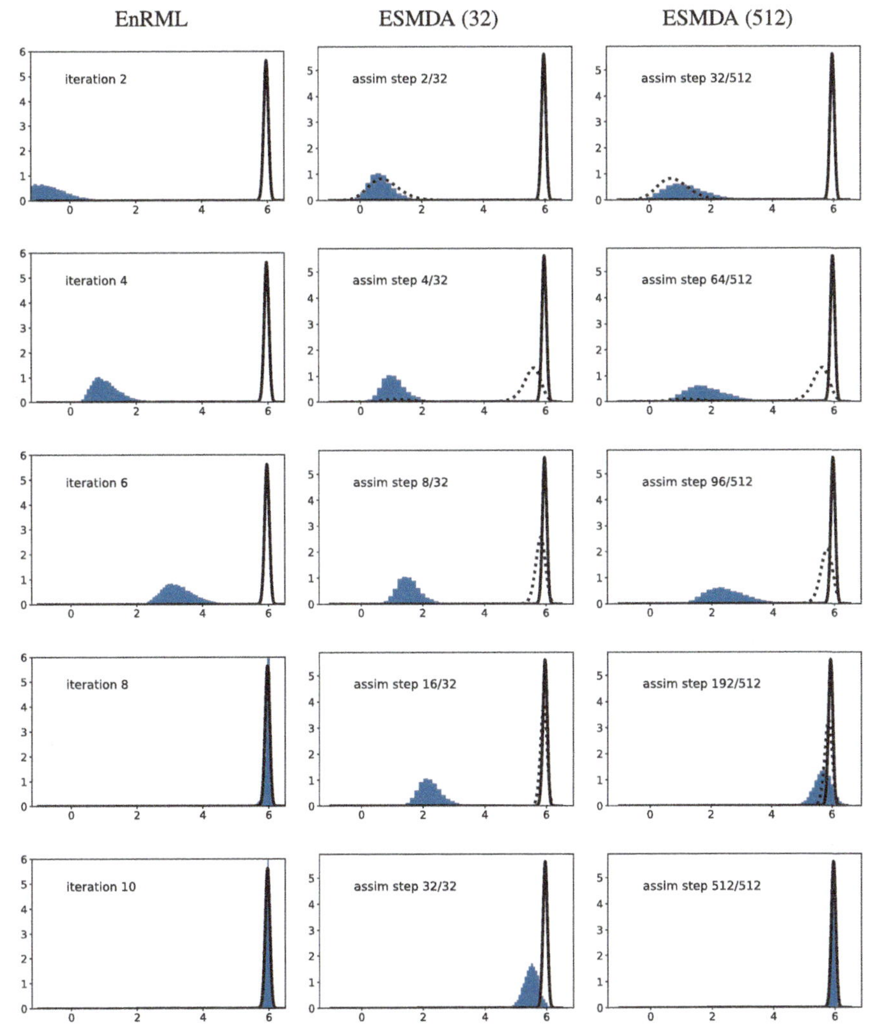

Fig. 9.6 Ensemble-based data assimilation for a moderately nonlinear monotonic observation function $g_1(x)$

algorithm samples the posterior distribution nearly perfectly after 8 to 10 iterations. In contrast, the ESMDA algorithm with 32 assimilation steps converges to a distribution quite far from the correct posterior distribution. The final distribution for ESMDA is improved when using 128 equally spaced data assimilation steps and further enhanced with 256 data assimilation steps. Increasing the number of assimilation steps beyond 256 results in minimal change to the final distribution (Fig. 9.5).

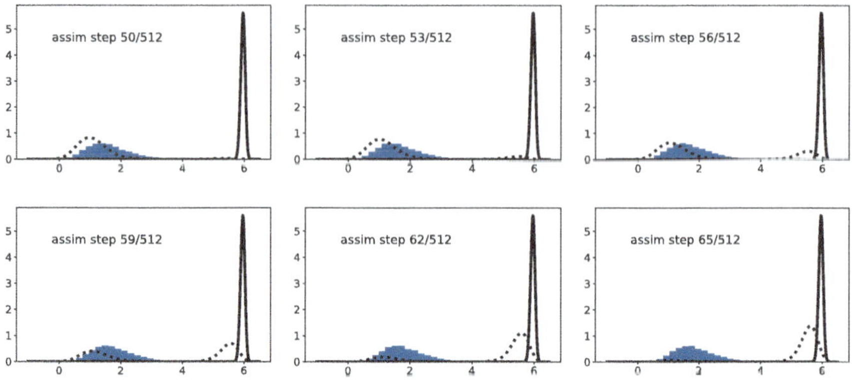

Fig. 9.7 Development of multi-modality at intermediate data assimilation steps for the moderately nonlinear observation function (g_1)

Ensemble data assimilation methods often work well for sampling from noisy distributions with a global maximum and many minor maxima. However, the methods usually perform poorly when multiple global modes of similar depth or height exist. For the observation function g_1, the failure of ESMDA to sample correctly results from the development of bimodal distributions with approximately equal mass in both modes at intermediate data assimilation steps. We can more easily see the growth and decline of the bimodal distributions for MDA in Fig. 9.6. For ESMDA_32, the bimodality between data assimilation steps 2 and 4 is apparent. In the earliest steps, the ensemble tracks the leftmost mode well, but with subsequent assimilation, the rightmost mode near $x = 5.5$ becomes predominant, but this mode has no samples. As the assimilation proceeds, the ensemble remains close to the first mode, which no longer has significant probability mass, and fails to sample the second mode despite the increasing importance of the second mode.

We see the behavior more clearly when using ESMDA with 512 data assimilation steps (Fig. 9.7). At MDA step 50, almost all probability mass should be located near $x = 1$, while we find that the ensemble of particles is slightly displaced to the right. At assimilation step 53, a small second mode at $x \approx 5.5$ appears. At steps 56 and 59, the mass in the second mode has increased, but the ensemble of particles remains at the leftmost mode. By step 65, almost all of the mass should be located at the rightmost mode as the leftmost mode is essentially gone. However, the ensemble of realizations has still not moved.

In the case with moderate nonlinearity, the prior probability density is unimodal, and the posterior probability density appears unimodal. Bimodality only appears at intermediate data assimilation steps. From Eqs. (2.4) and (3.18), it is straightforward to show that the target probability density at the kth data assimilation step in ESMDA is

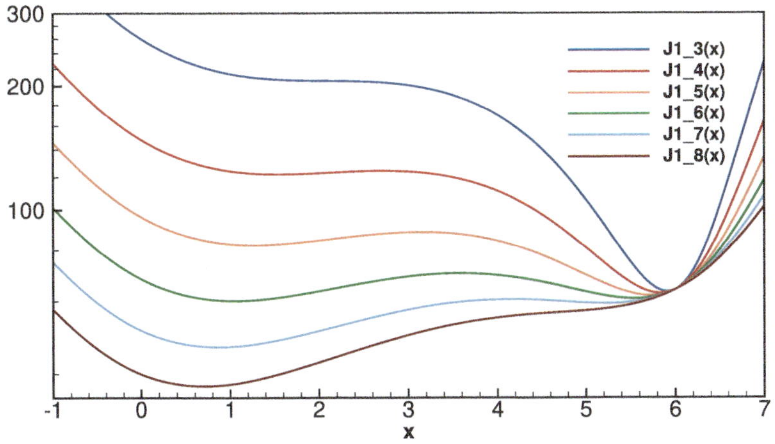

Fig. 9.8 The ESMDA inflation of measurement errors may lead to multimodal distributions even in the case with a monotonic measurement functional $g_1(x)$. Here, the cost functions with measurement standard deviation between four and seven show multimodal behavior

$$f_k(\mathbf{z}|\mathbf{d}) \propto \exp\Bigg[-\frac{1}{2}\left(\mathbf{z}-\mathbf{z}^{\mathrm{f}}\right)^{\mathrm{T}}\mathbf{C}_{zz}^{-1}\left(\mathbf{z}-\mathbf{z}^{\mathrm{f}}\right) \\ -\frac{1}{2}\big(\mathbf{g}(\mathbf{z})-\mathbf{d}\big)^{\mathrm{T}}\left(\sum_{i=1}^{k}\frac{1}{\alpha}\right)\mathbf{C}_{dd}^{-1}\big(\mathbf{g}(\mathbf{z})-\mathbf{d}\big)\Bigg]. \tag{9.6}$$

Thus, at the kth assimilation step in ESMDA with equal steps, we should generate samples from a distribution with measurement error variance inflated by α/k. Figure 9.8 shows the cost functions for the data assimilation problem using the moderately nonlinear observation function $g_1(x)$ at various levels of observation error standard deviation. Although several curves show bimodality in the objective function, affecting the ensemble evolution, only the green curve corresponding to a standard deviation of the measurement error of 6 (variance of 36) would be visibly bimodal in probability density. If we use ESMDA with 512 data assimilation steps, then the bimodality should be visible at step 62 (e.g., when $512/k = 36/4$). This result is verified in Fig. 9.7.

9.6 Assimilation of a Highly Nonlinear Measurement

For the highly nonlinear, non-monotonic observation operator, the posterior probability density is bimodal—although only noticeable in the cost function—and we do not expect ESMDA or EnRML to sample correctly or even provide a realistic uncertainty quantification. Results in Fig. 9.9 confirm the expectation of poor perfor-

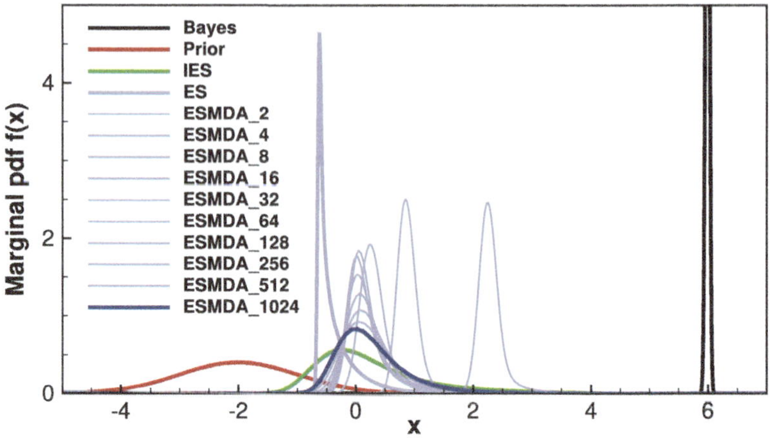

Fig. 9.9 Posterior estimate using the highly nonlinear function $g_2(x)$. In this case, both the EnRML and ESMDA solutions get trapped in the local minima around $x = 0$. Interestingly, the rightmost ESMDA solution is the ESMDA with two steps, and with an increasing number of steps the method converges to the local minimum

mance. The final distributions of particles for both methods center near the location of a local minimum in the cost function.

Both the ESMDA and the EnRML methods use ensemble-based gradients to compute the direction for model improvement in an iterative data assimilation step. In both cases, we base the estimates of data sensitivity to model parameters on the *current* ensemble of predicted data residuals. The two approaches differ, however, in their use of the ensemble of model mismatch realizations. The EnRML method estimates the prior covariance based on the *initial* ensemble of model parameters. In contrast, the ESMDA method uses the *current* ensemble of model parameters (the posterior from the previous assimilation step). If the observation operator is monotonic as it was for operators g_1 and g_2, the EnRML method generally converges to a good approximation of the posterior pdf. When the observation operator is non-monotonic, as it is for g_2, the updating direction for the ensemble members may be incorrect, as illustrated in Fig. 9.10. In this case, the mode at $x = 1.5$ is so tiny that we cannot see it in the figures, yet it affects the sampling. The sensitivity of the observation to the model variable is negative for g_2 in the approximate interval $0.5 < x < 3.5$, so the EnRML cannot advance the ensemble of realizations to the primary mode near $x = 6$. Both assimilation methods fail for this level of nonlinearity (Fig. 9.7).

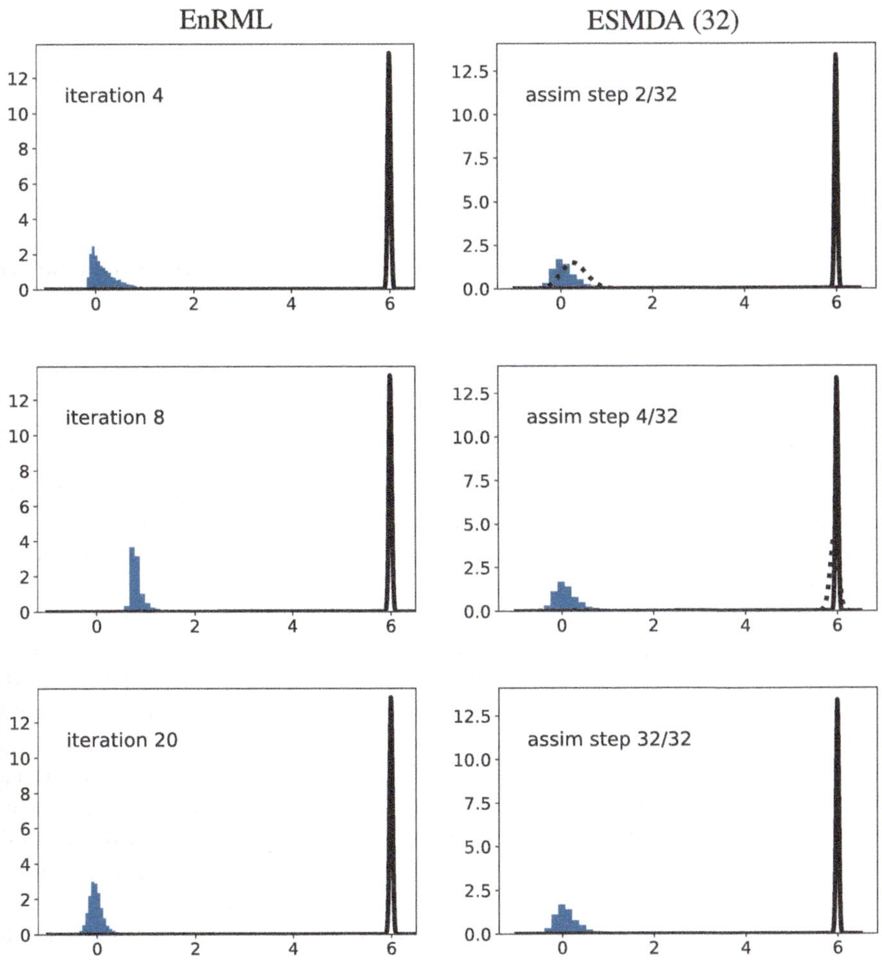

Fig. 9.10 Ensemble-based data assimilation for the highly nonlinear observation function (g_2)

9.7 Summary

One challenge with using ESMDA for history matching is that, as it is a recursive algorithm, the current assimilation update's quality depends on the previous update's quality. If an update is poor at some step, it may be difficult for the method to recover. Even for moderate nonlinearity (a monotonic observation function), intermediate probability densities can be bimodal, even though both the posterior and the prior pdfs appear unimodal. On the other hand, ESMDA is easy to apply; we can start with a single ES step for early testing of the parameterization of the prior and likelihood functions and later increase n_α to an affordable number of steps dependent on the size of the model and the available computer resources. Typically, when we history-

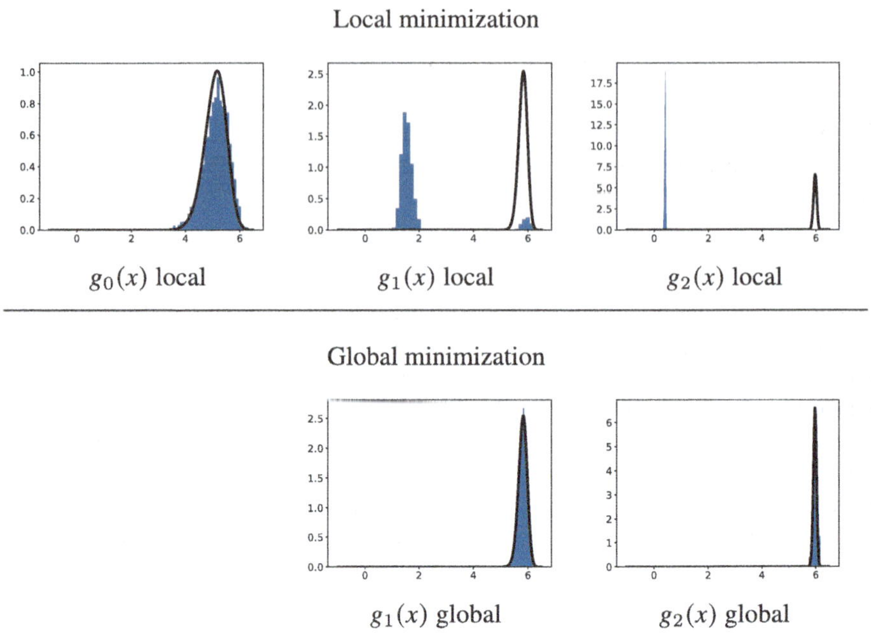

Fig. 9.11 RML for sampling with all three observation operators using both local and global optimization for minimization of the objective function

match reservoir models, we see a significant improvement in the results when using up to four or eight steps. Still, such a few steps may not suffice in highly nonlinear reservoir models.

Figure 9.11 illustrates the challenges introduced when using EnRML in the presence of minor local minima. Suppose we initiate the minimization with the sample from the prior as the initial guess, and we use a greedy minimization algorithm. In that case, the distribution of samples can be poor, and we may miss some modes entirely. Conversely, when using a global minimization, the sampling will generally not be correct, but there is a lesser tendency to get trapped in local minima. Contrary to ESMDA, EnRML is an iterative method, and the number of required iterations for convergence will depend on the degree of model nonlinearity.

There are only a few comparisons of ESMDA with EnRML on reservoir history-matching problems. Chen and Oliver (2013) compared ESMDA and EnRML on the Brugge benchmark history-matching problem. For this problem, the differences in results for the two data assimilation methods were relatively small. Still, as they did not use facies models for updating the permeability or porosity fields, the nonlinearity may not have been significant. Chen and Oliver (2013) compared ESMDA and EnRML on the small porous flow problem of Emerick and Reynolds (2013), and again, the differences were not significant.

The examples illustrate the importance of analyzing and understanding the properties of the minimization problem. It is possible to define poorly posed history-matching problems unless we are careful with the definitions of the prior statistics. For example, we would do much better for the examples discussed here with a prior distribution located closer to the global minimum to the right of $x = 3.5$. Thus, we have designed the current example to illustrate how badly things can go rather than provide a general comparison of the data assimilation methods used.

Part II
Robust Optimization and Closed-Loop Reservoir Management

Closed loop reservoir management and robust optimization build on and extend the reservoir-HM methods from Part I to also include the optimization of future controls with uncertainty and provide the reservoir engineer with tools that support the decision-making.

Chapter 10
Ensemble Optimization Method

Abstract The output of ensemble history-matching methods is an ensemble of model realizations approximating the posterior distribution conditioned on historical data. The goal is to use this ensemble representation of the reservoir uncertainty when planning future production and drainage strategies. In closed-loop reservoir management workflows, robust optimization methods are now a key component. These methods involve optimizing the expected value of future controls and production strategies over the ensemble of history-matched models. The "robustness" results from the explicit accounting for the reservoir uncertainty in the optimization. In this chapter, we will discuss a robust optimization method, EnOpt, where we define the optimum as the controls and strategies that result in the highest net present value when averaged over the ensemble of reservoir models.

10.1 Overview

The goal of a history-matching workflow is normally to create a reliable reservoir management model that we can use to perform"what-if" experiments, e.g., how much additional oil can we produce if we drill a new injector or an infill well in a given location?

Models calibrated to reproduce historical production data are generally more reliable at making predictions of quantities of interest than those that cannot reproduce past observations. Unfortunately, the available data is insufficient to specify a realistic model's properties uniquely or to provide unique predictions of the outcomes of contemplated actions. Instead, one potentially has many model parameter realizations that are equally good at matching the production data. When using an ensemble method for history matching, one typically obtains $\mathcal{O}(100)$ history-matched models available for reservoir production optimization. In this case, instead of making decisions based on a single best model or on low, median, and high models, the model ensemble offers the possibility of computing an expected outcome, which we can optimize. We use $\boldsymbol{\theta} \in \Re^n$ to represent the uncertain variables characterizing the geological model and a set of control variables $\mathbf{u} \in \Re^{\nu}$ defining the reservoir's

G. Evensen et al., *Ensemble History Matching*,
https://doi.org/10.1007/978-3-031-99155-4_10

development and operation. We represent the reservoir uncertainty by an ensemble of model states $\boldsymbol{\theta}_j$, where j is a counter over $j = 1, \ldots, N$ realizations.

Suppose we wish to find an optimal control vector $\mathbf{u}$ that leads to the highest expected value of an objective function $\Omega(\mathbf{u}, \boldsymbol{\theta}_j)$ (e.g., profitability). We approximate the expected value by averaging over an ensemble of reservoir model realizations, each of which is an independent draw from the posterior distribution, conditioned on production data,

$$\overline{\Omega(\mathbf{u}, \boldsymbol{\theta}_j)} = \frac{1}{N} \sum_{j=1}^{N} \Omega(\mathbf{u}, \boldsymbol{\theta}_j). \tag{10.1}$$

In Eq. (10.1), the average objective function over the ensemble of geological models $\boldsymbol{\theta}_j$ is computed for a given set of controls $\mathbf{u}$. In the ensemble-average operator, we always add a subscript j to the random variables used to approximate the expected value of the objective. A common goal of reservoir optimization is to find the controls $\mathbf{u}$ that maximize $\overline{\Omega(\mathbf{u}, \boldsymbol{\theta}_j)}$ (Chen et al. 2009; Chang et al. 2020 see, e.g.,). Common specific objectives are to maximize the economic value of a reservoir (or, more precisely, the net present value) by modifying the location of wells or the rates at which fluids are produced or injected.

The problem of maximizing the expected value of a reservoir management or field development objective is complex. Evaluating the objective for even a single control-variable setting on a single reservoir model parameter is computationally expensive as it requires running the reservoir simulator over the optimization period. Another challenge is that the sample average representation of the expected value may need too many model realizations for reasonable accuracy, in which case the number of function evaluations increases substantially. Finally, the derivatives of the objective with respect to the control variables may not be available, so convergence to the optimum may be slow.

When the objective function for reservoir optimization is differentiable with respect to the control variables, the most efficient optimization methods are generally gradient-based, even when the objective function's landscape has multiple minima. For most reservoir management or field development problems, the objective function is the output of a numerical reservoir flow simulator. Derivatives of the objective function are sometimes, but not always, available. When derivatives are unavailable, developing code for the computation of derivatives is occasionally feasible. Oliver et al. (2008, Sect. 9.7) provide a relatively straightforward and thorough description of the adjoint method for computing gradients for multi-phase flow but see also the more general overview, not specific to reservoir simulation, by Griewank (2003). The adjoint approach to the computation of derivatives for reservoir management has been utilized successfully (Nævdal et al. 2006; Wang et al. 2009 e.g.,), although generally not for optimization under uncertainty, as in that case evaluation of the gradient of Eq. (10.1) would require N solutions of the adjoint system at each iteration of the minimization.

An alternative, which does not require the solution of the adjoint system, is to use derivative-free optimization (DFO) methods. A DFO method uses only function evaluations for the optimization. Although methods such as particle swarm and evolutionary algorithms are also "derivative-free," the term DFO refers only to methods for local optimization of continuous-valued, single-objective problems (Larson et al. 2019; Moré and Wild 2009).

The EnOpt method provides a means to simplify the gradient computation in robust optimization. We can then compute one common stochastic gradient over the ensemble of geological realizations instead of treating each model realization independently. In the following section, we will introduce the EnOpt method for the case with a single model realization.

10.2 EnOpt Without Geological Uncertainty

We will first illustrate the EnOpt method for maximizing the objective function in the case without geological uncertainty, i.e., $N = 1$ in Eq. (10.1), and we only have one geological realization $\boldsymbol{\theta}$.

The starting point is the pre-conditioned steepest-ascent method,

$$\mathbf{u}^{i+1} = \mathbf{u}^{i} + \eta^{i}\, \mathbf{C}_{uu}\, \nabla_{\mathbf{u}}\, \Omega\big(\mathbf{u}^{i}, \boldsymbol{\theta}\big), \tag{10.2}$$

where the superscript i is the iteration index, η^{i} is the step size, $\mathbf{C}_{uu} \in \Re^{\nu\times\nu}$ is a pre-determined symmetric and positive-definite matrix, and $\nabla_{\mathbf{u}}\, \Omega(\mathbf{u}, \boldsymbol{\theta}) \in \Re^{\nu\times 1}$ is the gradient of the objective function to the control variables evaluated at the current values $(\mathbf{u}^{i}, \boldsymbol{\theta})$.

The EnOpt algorithm (Lorentzen et al. 2006; Chen et al. 2009; Chang et al. 2020) introduces an approximate stochastic representation of the gradient. The method computes the gradient from an ensemble of controls, $\mathbf{u}_j \sim \mathcal{N}(\mathbf{u}, \mathbf{C}_{uu})$, as a linear regression between the controls and the resulting NPVs. We use the subscript j as an index for a realization from the ensemble of control values. From here onwards, we only discuss the gradient approximation at the current iteration i. For the simplicity of notation, we skip the iteration superscript on $\mathbf{u}^{i}_{j}$ and note that the geological model is independent of i. For a given geological model $\boldsymbol{\theta}$, we then get an approximation of the gradient as

$$\begin{aligned}
\nabla_{\mathbf{u}}\, \Omega(\mathbf{u}, \boldsymbol{\theta}) &\triangleq \mathbf{C}_{uu}^{-1}\mathbf{C}_{uo} \\
&\approx \mathbf{C}_{uu}^{-1}\, \frac{1}{N} \sum_{j=1}^{N} \big(\mathbf{u}_j - \overline{\mathbf{u}}\big) \Big(\Omega\big(\mathbf{u}_j, \boldsymbol{\theta}\big) - \overline{\Omega\big(\mathbf{u}_j, \boldsymbol{\theta}\big)}\Big).
\end{aligned} \tag{10.3}$$

Here, $\overline{\mathbf{u}}$ is the arithmetic ensemble average of the controls at iteration i, and

$$\overline{\Omega\big(\mathbf{u}_j, \boldsymbol{\theta}\big)} = \frac{1}{N} \sum_{j=1}^{N} \Omega\big(\mathbf{u}_j, \boldsymbol{\theta}\big), \tag{10.4}$$

is the arithmetic average of the ensemble of NPVs obtained by evaluating the cost function for all control realizations. Thus, we only need to compute the sample covariance, $\mathbf{C}_{uo} \in \Re^{\upsilon \times 1}$, between the ensemble of controls $\mathbf{u}_j$ and the resulting ensemble of NPVs, $\Omega(\mathbf{u}_j, \boldsymbol{\theta})$, where we have retained the j subscript to indicate that the averaging is over the ensemble of controls. The EnOpt iteration then becomes

Standard EnOpt without geological uncertainty

$$\mathbf{u}^{i+1} = \mathbf{u}^i + \eta^i \frac{1}{N} \sum_{j=1}^{N} (\mathbf{u}_j - \overline{\mathbf{u}}) \left(\Omega(\mathbf{u}_j, \boldsymbol{\theta}) - \overline{\Omega(\mathbf{u}_j, \boldsymbol{\theta})}\right). \tag{10.5}$$

This formulation defines the original EnOpt algorithm as proposed by Lorentzen et al. (2006) and Chen et al. (2009). Note that we define the solution as *the optimal control vector defined by the ensemble average of the updated control vectors.*

10.3 EnOpt with Geological Uncertainty

When the reservoir properties are uncertain, it is reasonable to seek the optimal control vector that maximizes the expected value of the objective function. We have taken this approach when defining our objective function in Eq. (10.1).

A direct application of an ensemble of controls, as in the original EnOpt (Lorentzen et al. 2006), to a problem with geological uncertainty requires N simulations for each of the N geological realizations, and the method would become overwhelmingly computationally expensive. Chen et al. (2009) developed a version of EnOpt for cases where the reservoir properties and the optimal controls were uncertain.

Since we wish to compute a single control vector rather than an ensemble of control vectors, we only need one common gradient for all the geological realizations. Thus, it makes sense to use a Monte Carlo approach where we sample pairs of realizations $(\mathbf{u}_j, \boldsymbol{\theta}_j)$ from their representative distributions, as this simplifies the formulation, and we avoid running N^2 simulations. Stordal et al. (2016) showed that this approximation does not introduce any bias in the gradient approximation for a sufficiently large ensemble size. Thus, in the case of geological uncertainty, we can use the following definition for the gradient

$$\begin{aligned} \nabla_{\mathbf{u}} \Omega(\mathbf{u}, \boldsymbol{\theta}) &\triangleq \mathbf{C}_{uu}^{-1} \mathbf{C}_{uo} \\ &\approx \mathbf{C}_{uu}^{-1} \frac{1}{N} \sum_{j=1}^{N} (\mathbf{u}_j - \overline{\mathbf{u}}) \left(\Omega(\mathbf{u}_j, \boldsymbol{\theta}_j) - \overline{\Omega(\mathbf{u}_j, \boldsymbol{\theta}_j)}\right). \end{aligned} \tag{10.6}$$

Here we have extended the ensemble-average definition from Eq. (10.4) to also include the ensemble of geological realizations as

$$\overline{\Omega(\mathbf{u}_j, \boldsymbol{\theta}_j)} = \frac{1}{N} \sum_{j=1}^{N} \Omega(\mathbf{u}_j, \boldsymbol{\theta}_j). \tag{10.7}$$

The standard EnOpt algorithm for an ensemble of reservoir models then becomes

Standard EnOpt with geological uncertainty

$$\mathbf{u}^{i+1} = \mathbf{u}^i + \eta^i \frac{1}{N} \sum_{j=1}^{N} \left(\mathbf{u}_j - \overline{\mathbf{u}}\right) \left(\Omega(\mathbf{u}_j, \boldsymbol{\theta}_j) - \overline{\Omega(\mathbf{u}_j, \boldsymbol{\theta}_j)}\right). \tag{10.8}$$

A further modification of the EnOpt algorithm with geological uncertainty, introduced by Fonseca (2016), modifies the averaging of the NPVs in Eq. (10.7) by using the following definition for the gradient

$$\nabla_{\mathbf{u}} \Omega(\mathbf{u}, \boldsymbol{\theta}) \approx \mathbf{C}_{uu}^{-1} \frac{1}{N} \sum_{j=1}^{N} \left(\mathbf{u}_j - \overline{\mathbf{u}}\right) \left(\Omega(\mathbf{u}_j, \boldsymbol{\theta}_j) - \Omega(\overline{\mathbf{u}}, \boldsymbol{\theta}_j)\right). \tag{10.9}$$

In cases where the geological uncertainty is high, Eq. (10.9) gives a lower variance for the gradient estimate than Eq. (10.6), see Stordal et al. (2016). Therefore, this work will use the EnOpt method with Eq. (10.9). In this version of EnOpt, it is necessary to compute the NPV using $\overline{\mathbf{u}}$ for all geological realizations in each iteration. Thus, each optimization iteration requires $2N$ simulations.

StoSAG EnOpt with geological uncertainty

$$\mathbf{u}^{i+1} = \mathbf{u}^i + \eta^i \frac{1}{N} \sum_{j=1}^{N} \left(\mathbf{u}_j - \overline{\mathbf{u}}\right) \left(\Omega(\mathbf{u}_j, \boldsymbol{\theta}_j) - \Omega(\overline{\mathbf{u}}, \boldsymbol{\theta}_j)\right). \tag{10.10}$$

10.4 Optimization Using the Mean Model

A substantial geological uncertainty will lead to a more noisy gradient and reduce the covariance between the controls and the resulting NPVs. The recent mean-model bias correction (MMBC) method by Wang and Oliver (2021) tries to resolve this issue by optimizing the controls on the mean model (defined as the ensemble mean of the geological realizations). It is then possible to use the EnOpt method (formulated without geological uncertainty above) on the mean model and find the optimal solution more efficiently.

Let us start by writing the objective function in Eq. (10.1) as

$$\begin{aligned}\overline{\Omega(\mathbf{u}, \boldsymbol{\theta}_j)} &= \frac{1}{N}\sum_{j=1}^{N} \frac{\Omega(\mathbf{u}, \boldsymbol{\theta}_j)}{\Omega(\mathbf{u}, \overline{\boldsymbol{\theta}})}\,\Omega(\mathbf{u}, \overline{\boldsymbol{\theta}}) \\ &= \alpha(\mathbf{u})\,\Omega(\mathbf{u}, \overline{\boldsymbol{\theta}}),\end{aligned} \tag{10.11}$$

where the precise value of α is

$$\alpha(\mathbf{u}) = \frac{1}{N}\sum_{j=1}^{N} \frac{\Omega(\mathbf{u}, \boldsymbol{\theta}_j)}{\Omega(\mathbf{u}, \overline{\boldsymbol{\theta}})}. \tag{10.12}$$

Hence, to compute $\alpha(\mathbf{u})$ for a particular value of $\mathbf{u}$, we need to run $N + 1$ model simulations, and if we have an ensemble of N control vectors $\mathbf{u}_j$ the cost is $N(N + 1)$ model simulations.

For a given control, $\mathbf{u}$, we have from Eq. (10.11) that generally $\Omega(\mathbf{u}, \overline{\boldsymbol{\theta}}) \neq \overline{\Omega(\mathbf{u}, \boldsymbol{\theta}_j)}$. If we set $\alpha = 1$, we will minimize a different objective function than the original EnOpt one. To reduce this bias, we will estimate a value of α that improves the objective function approximation.

The simplest approximation for α is to just set it to one. This choice leads to the most efficient method as we have no geological uncertainty impacting the gradient computations, and we need only to compute N simulations per iteration. However, we are solving a problem different from the original one, but with low geological uncertainty, this approach will still often provide an excellent optimal control vector. In the next chapter we will elaborate on the optimization using the mean model and introduce an additive bias correction method that makes the mean-model approach more efficient.

Chapter 11
Mean-Model Bias Correction Method

Abstract Optimization using the EnOpt method discussed in the previous chapter can be expensive for realistic field models. We are therefore encouraged to look for less computationally costly alternatives that reduce the number of required reservoir simulations. One such approach is to optimize the controls using the mean reservoir model, as was discussed in Sect. 10.4, rather than optimizing over the whole ensemble of geological models. When the geologic uncertainty is small, or the modeled response to changes in the controls is weakly nonlinear, this approach may give reasonable results. However, in many realistic cases, optimizing the objective function over the whole model ensemble will provide a different result than the solution obtained when we optimize using the mean model. This chapter will delve into an approach designed to reduce the bias in the approximate objective function, which is a crucial step in ensuring the accuracy of our optimization process.

11.1 Introduction

Wang and Oliver (2021) introduced the Mean-Model Bias Correction (MMBC) method as a more efficient alternative to EnOpt. Evensen and Chang (2022) explained the method further using a unified formulation for EnOpt and MMBC with a multiplicative bias correction term. In this section, we will revise the MMBC method using an *additive* representation for the bias correction described by Oliver (2024) rather than the multiplicative one from Sect. 10.4.

The computational cost of minimizing the expected value of the objective function in Eq. (10.1) can be infeasibly high for reservoir production optimization problems when the cost of evaluating the individual objective functions is high. In standard approaches to robust optimization, one either needs to evaluate all individual objective functions at each iteration, i.e., $\Omega(\mathbf{u}, \boldsymbol{\theta}_j)$ for all $j \in [1, \ldots, N]$, or one uses a noisy estimate of the gradient which reduces the improvement that can be achieved in an iteration. In the first case, each iteration is expensive, in the second case more iterations are required for convergence.

It would be ideal if, instead of evaluating the ensemble of objective functions at each of the model realizations $\boldsymbol{\theta}_j$, a single surrogate reservoir model, $\widehat{\boldsymbol{\theta}}$, could be used

G. Evensen et al., *Ensemble History Matching*,
https://doi.org/10.1007/978-3-031-99155-4_11

for decision making, specifically one that resulted in the same optimal decisions as would be made using the entire ensemble of model realizations. In that case, we could replace the ensemble of models with the model $\widehat{\boldsymbol{\theta}}$ such that the objective in Eq. (10.1) becomes

$$\overline{\Omega(\mathbf{u}, \boldsymbol{\theta}_j)} = \frac{1}{N} \sum_{j=1}^{N} \Omega(\mathbf{u}, \boldsymbol{\theta}_j) \approx \Omega(\mathbf{u}, \widehat{\boldsymbol{\theta}}), \tag{11.1}$$

where $\mathbf{u}$ is the vector of controls (e.g., well location, injection rate, or drilling order) and Ω is the objective to be optimized, e.g., net present value (NPV) or greenhouse gas (GHG) emissions. If the approximation is good, then one might be able to perform the optimization using only the surrogate model, i.e.,

$$\mathbf{u}^{\text{opt}} = \operatorname{argmin} \Omega(\mathbf{u}, \widehat{\boldsymbol{\theta}}). \tag{11.2}$$

The "mean model" has sometimes been used as a surrogate for the ensemble in reservoir production optimization (Brouwer et al. 2004; Nævdal et al. 2006; Wang et al. 2009; Chen et al. 2010), although the justification has generally been simply that the sample average approximation of the expectation in Eq. (10.4) is too expensive to use for optimization. In certain situations, however, the mean model is a rigorous replacement for the sample average approximation, and the "approximation" in Eq. (11.1) is exact. If, for example, the objective function is linear in the uncertain parameters of the model, such that

$$E\left[\Omega^{\text{lin}}(\mathbf{u}, \boldsymbol{\theta})\right] = \Omega^{\text{lin}}(\mathbf{u}, \overline{\boldsymbol{\theta}}), \tag{11.3}$$

then the correct choice is to use the mean model $\overline{\boldsymbol{\theta}}$ for $\widehat{\boldsymbol{\theta}}$. As an example, if the only uncertain "model parameter" is the oil price, then, as the NPV is a linear function of oil price, it is valid to use the mean oil price schedule instead of an ensemble of realizations of oil price schedules for optimization (Siraj et al. 2015).

In general, however, the relationship between quantities like NPV and reservoir permeability is nonlinear, so using the mean model for optimization does not result in accurate optimal controls. In that case, we can add a bias correction term that is a function of the control variables to the objective function evaluated at the mean model, i.e.,

$$\overline{\Omega(\mathbf{u}, \boldsymbol{\theta}_j)} = \frac{1}{N} \sum_{j=1}^{N} \Omega(\mathbf{u}, \boldsymbol{\theta}_j) = \Omega(\mathbf{u}, \widehat{\boldsymbol{\theta}}) + \alpha(\mathbf{u}), \tag{11.4}$$

where the correct value for $\alpha(\mathbf{u})$ is

$$\alpha(\mathbf{u}) = \overline{\Omega(\mathbf{u}, \boldsymbol{\theta}_j)} - \Omega(\mathbf{u}, \widehat{\boldsymbol{\theta}}). \tag{11.5}$$

Although Wang and Oliver (2021) and Evensen and Chang (2022) used a multiplicative bias correction as discussed in Sect. 10.4, we present an additive bias correction here because it is more robust in the case of small values of the objective function (Oliver 2024).

11.2 Bias Correction

From this point on, we will refer to the surrogate model $\widehat{\boldsymbol{\theta}}$ as the mean model $\overline{\boldsymbol{\theta}}$ and will defer until later the specifics of the definition of the mean model. The bias correction at control setting $\mathbf{u}$ is defined to be the difference between the objective function obtained from the evaluation using the mean model and the sample average approximation of the objective function obtained from evaluation using an ensemble of model realizations

$$\begin{aligned}\alpha(\mathbf{u}) &= \overline{\Omega\big(\mathbf{u}, \boldsymbol{\theta}_j\big)} - \Omega\big(\mathbf{u}, \overline{\boldsymbol{\theta}}\big) \\ &= \frac{1}{N}\sum_{j=1}^{N}\Big[\Omega\big(\mathbf{u}, \boldsymbol{\theta}_j\big) - \Omega\big(\mathbf{u}, \overline{\boldsymbol{\theta}}\big)\Big] \\ &= \frac{1}{N}\sum_{j=1}^{N} b_j\big(\mathbf{u}, \boldsymbol{\theta}_j\big),\end{aligned} \tag{11.6}$$

and we refer to the difference between the value of the objective function when applied to a single model realization and to the mean model,

$$b_j(\mathbf{u}, \boldsymbol{\theta}_j) = \Omega\big(\mathbf{u}, \boldsymbol{\theta}_j\big) - \Omega\big(\mathbf{u}, \overline{\boldsymbol{\theta}}\big), \tag{11.7}$$

as the partial additive bias correction at control setting $\mathbf{u}$ for model realization $\boldsymbol{\theta}_j$.

In most cases of interest, it is neither necessary nor possible to fully characterize the function $b_j(\mathbf{u})$. Instead, we are limited to evaluating and estimating the partial correction factor at a finite number of control realizations, $\mathbf{u} \in (\mathbf{u}_1, \ldots \mathbf{u}_{N_u})$. In some production optimization problems, we only define $\mathbf{u}$ at a discrete set of values. For example, in a well-location problem, we may attempt to locate wells in an optimal grid cell. In other cases, such as rate optimization, the set $\mathbf{u} \in (\mathbf{u}_1, \ldots \mathbf{u}_{N_u})$ includes only evaluated control-variable settings. Note that Table 11.1 summarizes the dimensions used for easy reference.

We define the vector of all partial correction factors $\mathbf{b}_j$ for model realization $\boldsymbol{\theta}_j$ and the control variables vectors $\mathbf{u}_i \in (\mathbf{u}_1, \ldots \mathbf{u}_{N_u})$, as

Table 11.1 The table shows the notation used for model state and control vectors, with their dimensions and ensemble sizes

ν	Dimension of the control vector $\mathbf{u}$
N_u	Ensemble size of control vectors $\mathbf{u}_j$
n	State dimension of model state vector $\boldsymbol{\theta}$
N	Ensemble size of model state vectors $\boldsymbol{\theta}_j$

$$\mathbf{b}_j = \begin{bmatrix} \beta_{1j}(\mathbf{u}_1, \boldsymbol{\theta}_j) \\ \beta_{2j}(\mathbf{u}_2, \boldsymbol{\theta}_j) \\ \vdots \\ \beta_{N_u j}(\mathbf{u}_{N_u}, \boldsymbol{\theta}_j) \end{bmatrix}, \tag{11.8}$$

and similarly define the matrix $\mathbf{B} \in \Re^{N_u \times N}$ of all partial correction factors β_{ij} for model realization $\boldsymbol{\theta}_j$ for each possible control variables vector $\mathbf{u}_i$,

$$\mathbf{B} = \begin{bmatrix} \mathbf{b}_1 \ \mathbf{b}_2 \ \dots \ \mathbf{b}_N \end{bmatrix} = \begin{bmatrix} \beta_{11} & \beta_{12} & \dots & \beta_{1N} \\ \beta_{21} & \beta_{22} & \dots & \beta_{2N} \\ \vdots & \vdots & & \vdots \\ \beta_{N_u 1} & \beta_{N_u 2} & \dots & \beta_{N_u N} \end{bmatrix}. \tag{11.9}$$

The bias correction function, $\alpha(\mathbf{u}_i)$ for control setting $\mathbf{u}_i$, is the ensemble average of all partial bias corrections,

Estimation of the additive bias corrections $\alpha(\mathbf{u}_i)$

$$\alpha(\mathbf{u}_i) = \frac{1}{N} \sum_{j=1}^{N} b_j(\mathbf{u}_i, \boldsymbol{\theta}_j) \tag{11.10}$$

With this definition, we have a bias correction $\alpha(\mathbf{u}_i)$ for each realization of the control vector $\mathbf{u}_i$. In the following section, we will derive an approximate $\alpha(\mathbf{u})$ estimator given the computed bias corrections for the finite number of control settings realizations $\mathbf{u}_i$.

11.3 Bias Estimation

To estimate the bias correction at one control setting, from the "observation" of the bias correction at other control settings, we assume that the partial bias for model realization $\boldsymbol{\theta}_j$ is a continuous function of the "distance" between control variable settings and that we can model it as a Gaussian random field with stationary mean and covariance that depends only on the distance between the control variables. Thus, we assume

$$\mathbf{b}_j \sim \mathcal{N}\left(b_j^{\mathrm{pri}} \mathbf{1}_{N_u}, \mathbf{C}_{bb}\right), \tag{11.11}$$

with some prior estimate b_j^{pri}. Here, $\mathbf{1}_{N_u}$ is a column vector of length N_u, all of whose elements are equal to one, and $\mathbf{C}_{bb}$ is the assumed covariance matrix for the random variable $\mathbf{b}_j \in \Re^{N_u}$. Experience indicates that this assumption is reasonable for many types of controls, such as well-location or production rate.

We also assume that for each model realization, $\boldsymbol{\theta}_j$, we can model the mean over $\mathbf{u}$ of partial bias corrections (the elements b_j^{pri}) as Gaussian, but with uncertain mean and uncertain variance.

$$b_j^{\mathrm{pri}} \sim \mathcal{N}\left(\alpha^{\mathrm{pri}}, \sigma_b^2\right). \tag{11.12}$$

We will examine the reasonableness of this type of modeling in a simple well-location problem described in Sect. 11.7. As we seldom have insight into the mean value of the bias correction α before we "observe" its values, we provide α^{pri} with an uninformative (uniform) prior probability density.

We can now write the joint prior probability density for $\mathbf{B}$, b_j^{pri} and α^{pri} as

$$\begin{aligned} f\left(\mathbf{B}, b_j^{\mathrm{pri}}, \alpha^{\mathrm{pri}}\right) &\propto \exp\left(-\frac{1}{2}\sum_{j=1}^{N}\left(\mathbf{b}_j - b_j^{\mathrm{pri}}\mathbf{1}_{N_u}\right)^{\mathrm{T}} \mathbf{C}_{bb}^{-1}\left(\mathbf{b}_j - b_j^{\mathrm{pri}}\mathbf{1}_{N_u}\right)\right) \\ &\times \exp\left(-\frac{1}{2}\sum_{j=1}^{N}\left(b_j^{\mathrm{pri}} - \alpha^{\mathrm{pri}}\right)^2 / \sigma_b^2\right), \end{aligned} \tag{11.13}$$

and the likelihood function for measurements $\mathbf{d}$ of $\mathbf{B}$ as

$$f(\mathbf{d}|\mathbf{B}) \propto \exp\left(-\frac{1}{2}\sum_{j=1}^{N}\left(\mathbf{H}_j\mathbf{b}_j - \mathbf{d}_j\right)^{\mathrm{T}} \mathbf{C}_{dd}^{-1}\left(\mathbf{H}_j\mathbf{b}_j - \mathbf{d}_j\right)\right), \tag{11.14}$$

where $\mathbf{d}_j$ is a vector of samples of $\mathbf{b}_j$ evaluated at several control variables and $\mathbf{H}_j$ is a matrix whose rows are all zeros except for a value of one at the columns corresponding to observations of $\mathbf{b}_j$.

The posterior probability distribution function for $\mathbf{B}$, b_j^{pri} and α^{pri} conditional to the observations of elements of $\mathbf{B}$ is

$$
\begin{aligned}
f\left(\mathbf{B}, b_j^{\mathrm{pri}}, \alpha^{\mathrm{pri}} \mid \mathbf{d}\right) \propto & \exp\left(-\frac{1}{2}\sum_{j=1}^{N}\left(\mathbf{b}_j - b_j^{\mathrm{pri}}\mathbf{1}_{N_u}\right)^{\mathrm{T}}\mathbf{C}_{bb}^{-1}\left(\mathbf{b}_j - b_j^{\mathrm{pri}}\mathbf{1}_{N_u}\right)\right) \\
& \times \exp\left(-\frac{1}{2}\sum_{j=1}^{N}\left(b_j^{\mathrm{pri}} - \alpha^{\mathrm{pri}}\right)^2/\sigma_b^2\right) \\
& \times \exp\left(-\frac{1}{2}\sum_{j=1}^{N}\left(\mathbf{H}_j\mathbf{b}_j - \mathbf{d}_j\right)^{\mathrm{T}}\mathbf{C}_{dd}^{-1}\left(\mathbf{H}_j\mathbf{b}_j - \mathbf{d}_j\right)\right).
\end{aligned}
\tag{11.15}
$$

For some optimization problems, identifying a helpful measure of distance between control variables may not be straightforward. For the optimal scheduling of drilling, Wang and Oliver (2021) found that the bias correction was quite sensitive to the metric used to measure the distance between two sequences of drilling order.

Letting $\mathcal{S}(\mathbf{B}, b_j^{\mathrm{pri}}, \alpha^{\mathrm{pri}})$ denote the logarithm of the posterior probability density in Eq. (11.15), we obtain the maximum a posteriori estimates of $\mathbf{B}$, b_j^{pri}, and α^{pri} by solving for the location at which the gradient of the log-objective function vanishes. Thus, we have

$$
\nabla_{\mathbf{b}_j}\mathcal{S} = -\mathbf{C}_{bb}^{-1}\left(\mathbf{b}_j - b_j^{\mathrm{pri}}\mathbf{1}_{N_u}\right) - \mathbf{H}_j^{\mathrm{T}}\mathbf{C}_{dd}^{-1}\left(\mathbf{H}_j\mathbf{b}_j - \mathbf{d}_j\right) = 0, \tag{11.16}
$$

$$
\nabla_{b_j^{\mathrm{pri}}}\mathcal{S} = \mathbf{1}_{N_u}^{\mathrm{T}}\mathbf{C}_{bb}^{-1}\left(\mathbf{b}_j - b_j^{\mathrm{pri}}\mathbf{1}_{N_u}\right) - \left(b_j^{\mathrm{pri}} - \alpha^{\mathrm{pri}}\right)/\sigma_b^2 = 0, \tag{11.17}
$$

for $j = 1, \ldots, N$ and

$$
\nabla_{\alpha^{\mathrm{pri}}}\mathcal{S} = \alpha^{\mathrm{pri}} - \frac{1}{N}\sum_{j=1}^{N} b_j^{\mathrm{pri}} = 0. \tag{11.18}
$$

Solving Eq. (11.16) for $\mathbf{b}_j$, we obtain

$$
\begin{aligned}
\mathbf{b}_j = & \left(\mathbf{C}_{bb}^{-1} + \mathbf{H}_j^T\mathbf{C}_{dd}^{-1}\mathbf{H}_j\right)^{-1}\mathbf{C}_{bb}^{-1}\mathbf{1}_{N_u}b_j^{\mathrm{pri}} \\
& + \left(\mathbf{C}_{bb}^{-1} + \mathbf{H}_j^T\mathbf{C}_{dd}^{-1}\mathbf{H}_j\right)^{-1}\mathbf{H}_j^T\mathbf{C}_{dd}^{-1}\mathbf{d}_j.
\end{aligned}
\tag{11.19}
$$

Then, using the Woodbury-Sherman-Morrison identity, we obtain a smaller matrix inversion,

Estimation of the partial corrections $\mathbf{b}_j$

$$
\begin{aligned}
\mathbf{b}_j = & \left(\mathbf{I} - \mathbf{C}_{bb}\mathbf{H}^T\left(\mathbf{H}_j\mathbf{C}_{bb}\mathbf{H}_j^T + \mathbf{C}_{dd}\right)^{-1}\mathbf{H}_j\right)\mathbf{1}_{N_u}b_j^{\mathrm{pri}} \\
& + \mathbf{C}_{bb}\mathbf{H}_j^T\left(\mathbf{H}_j\mathbf{C}_{bb}\mathbf{H}_j^T + \mathbf{C}_{dd}\right)^{-1}\mathbf{d}_j.
\end{aligned}
\tag{11.20}
$$

From Eq. (11.17), we obtain a second equation for $\mathbf{b}_j$,

$$\mathbf{1}_{N_u}^T \mathbf{C}_{bb}^{-1} \mathbf{b}_j - \left(\mathbf{1}_{N_u}^T \mathbf{C}_{bb}^{-1} \mathbf{1}_{N_u} + \sigma_b^{-2}\right) b_j^{\text{pri}} + \sigma_b^{-2} \alpha^{\text{pri}} = 0, \tag{11.21}$$

which, together with Eq. (11.20), can be used to eliminate $\mathbf{b}_j$, resulting in an equation involving only the unknown hyperparameters b_j^{pri} and α^{pri},

$$\begin{aligned} \mathbf{1}_{N_u}^T \mathbf{C}_{bb}^{-1} \Big[&\left(\mathbf{I} - \mathbf{C}_{bb} \mathbf{H}_j^T \left(\mathbf{H}_j \mathbf{C}_{bb} \mathbf{H}_j^T + \mathbf{C}_{dd}\right)^{-1} \mathbf{H}_j \mathbf{1}_{N_u} b_j^{\text{pri}}\right. \\ &+ \mathbf{C}_{bb} \mathbf{H}_j^T \left(\mathbf{H}_j \mathbf{C}_{bb} \mathbf{H}_j^T + \mathbf{C}_{dd}\right)^{-1} \mathbf{d}_j \Big] \\ &- \left(\mathbf{1}_{N_u}^T \mathbf{C}_{bb}^{-1} \mathbf{1}_{N_u} + \sigma_b^{-2}\right) b_j^{\text{pri}} + \sigma_b^{-2} \alpha^{\text{pri}} = 0, \end{aligned} \tag{11.22}$$

or

$$\begin{aligned} &\left[-\mathbf{1}_{N_u}^T \mathbf{H}_j^T \left(\mathbf{H}_j \mathbf{C}_{bb} \mathbf{H}_j^T + \mathbf{C}_{dd}\right)^{-1} \mathbf{H}_j \mathbf{1}_{N_u} - \sigma_b^{-2} \right] b_j^{\text{pri}} \\ &\quad + \sigma_b^{-2} \alpha^{\text{pri}} + \mathbf{1}_{N_u}^T \mathbf{H}_j^T \left(\mathbf{H}_j \mathbf{C}_{bb} \mathbf{H}_j^T + \mathbf{C}_{dd}\right)^{-1} \mathbf{d}_j = 0. \end{aligned} \tag{11.23}$$

Note that $\mathbf{1}_{N_u}^T$ is a vector whose length is the number of control variables in the observed set and each of whose entries is 1. $\mathbf{H}_j$ is a matrix whose rows are all zeros except for one at the ith column for an observation of $b_j(x_i)$ so

$$\mathbf{H}_j \mathbf{1}_{N_u} = \mathbf{1}_j, \tag{11.24}$$

where the length of $\mathbf{1}_j$ is the number of observations of partial bias correction factors using model realization m_j. Similarly, we denote the product

$$\mathbf{H}_j \mathbf{C}_{bb} \mathbf{H}_j^T = \mathbf{C}_{bb}^{jj}. \tag{11.25}$$

We can then rewrite Eq. (11.23) in a simplified form as

$$\left[-\mathbf{1}_j^T \left(\mathbf{C}_{bb}^{jj} + \mathbf{C}_{dd}\right)^{-1} \mathbf{1}_j - \sigma_b^{-2} \right] b_j^{\text{pri}} + \sigma_b^{-2} \alpha^{\text{pri}} + \mathbf{1}_j^T \left(\mathbf{C}_{bb}^{jj} + \mathbf{C}_{dd}\right)^{-1} \mathbf{d}_j = 0, \tag{11.26}$$

and solve for b_j^{pri} to get

Estimation of the partial correction mean b_j^{pri}

$$b_j^{\text{pri}} = \frac{\sigma_b^{-2} \alpha^{\text{pri}} + \mathbf{1}_j^T \left(\mathbf{C}_{bb}^{jj} + \mathbf{C}_{dd}\right)^{-1} \mathbf{d}_j}{\mathbf{1}_j^T \left(\mathbf{C}_{bb}^{jj} + \mathbf{C}_{dd}\right)^{-1} \mathbf{1}_j + \sigma_b^{-2}}. \tag{11.27}$$

Then, by summing Eq. (11.27) over all j and applying the definition of α^{pri} in Eq. (11.18), we obtain

$$\alpha^{\mathrm{pri}} = \frac{1}{N}\sum_{j=1}^{N} \frac{\sigma_b^{-2}\alpha^{\mathrm{pri}} + \mathbf{1}_j^T\left(\mathbf{C}_{bb}^{jj} + \mathbf{C}_{dd}\right)^{-1}\mathbf{d}_j}{\mathbf{1}_j^T\left(\mathbf{C}_{bb}^{jj} + \mathbf{C}_{dd}\right)^{-1}\mathbf{1}_j + \sigma_b^{-2}}. \tag{11.28}$$

Finally, we obtain an expression that can be solved for the mean bias, α^{pri},

Estimation of the mean bias α^{pri}

$$\begin{aligned}\alpha^{\mathrm{pri}}\left(N - \sum_{j=1}^{N}\left(\mathbf{1}_j^T\left(\mathbf{C}_{bb}^{jj} + \mathbf{C}_{dd}\right)^{-1}\mathbf{1}_j\sigma_b^2 + 1\right)^{-1}\right) \\ = \sum_{j=1}^{N} \frac{\mathbf{1}_j^T\left(\mathbf{C}_{bb}^{jj} + \mathbf{C}_{dd}\right)^{-1}\mathbf{d}_j}{\mathbf{1}_j^T\left(\mathbf{C}_{bb}^{jj} + \mathbf{C}_{dd}\right)^{-1}\mathbf{1}_j + \sigma_b^{-2}}.\end{aligned} \tag{11.29}$$

To estimate the bias correction, one needs to first "observe" several values of the partial bias correction. These are the differences between the objective function values obtained using model realizations and those obtained using the mean model.

We can summarize the additive bias correction estimation as a sequence of relatively simple steps:

1. Estimate the hyperparameters σ_β, σ_b, and λ_b of the hierarchical distribution for the bias correction as described in Sect. 11.4.
2. Estimate the mean bias α^{pri} from the observations of partial corrections using Eq. (11.29).
3. Estimate the mean of the partial corrections for each of the model realizations, b_j^{pri}, using Eq. (11.27).
4. Estimate the partial corrections b_j for each of the model realizations $\boldsymbol{\theta}_j$ using Eq. (11.20).
5. Finally, estimate the additive bias correction $\alpha(\mathbf{u})$ from Eq. (11.10).

11.4 Hyperparameter Estimation

The prior distribution for $\mathbf{B}$, b_j^{pri}, and α^{pri} in Eq. (11.15) requires the specification of several hyperparameters, which will, in practice, be unknown. The covariance, $\mathbf{C}_{bb}$, specifies the degree to which a value of $\mathbf{b}_j(\mathbf{u})$ is related to a value of $\mathbf{b}_j(\mathbf{u}')$ and the variance σ_b^2 specifies the variability in the mean bias for individual model realizations. If the bias corrections at the two different control values are uncorrelated, then an observation of bias correction at $\mathbf{u}$ will not influence the estimate at $\mathbf{u}'$. Similarly, if the variance in the values is large, the estimates will have significant uncertainty. For bias correction estimation, we assume that $\mathbf{C}_{bb}$ is of the form

$$\mathbf{C}_{bb} = \sigma_\beta^2 \exp\left(-\frac{3(\mathbf{u}-\mathbf{u}')^\mathrm{T}(\mathbf{u}-\mathbf{u}')}{\lambda_b^2}\right). \tag{11.30}$$

In this case, it is only necessary to estimate the two hyperparameters σ_β and λ_b to define the covariance. Additionally, the prior distribution for the b_j^{pri}'s requires specification of the variance, σ_b^2.

We estimate σ_b^2 by averaging the variance estimates of the $b_j(\mathbf{u}_i, \boldsymbol{\theta}_j)$ from the initial evaluation points for fixed i. Furthermore, we estimate the variance, σ_β^2, of the partial bias correction factors as the mean-variance of function evaluations for the same model realization but different control variables. If we use the same model realizations at each initial control-variable evaluation location, then we can use multiple realizations of $\Omega(\mathbf{u}, \boldsymbol{\theta}_j) - \Omega(\mathbf{u}', \boldsymbol{\theta}_j)$ to estimate the magnitude of the correlation for $\mathbf{u} - \mathbf{u}'$. Fitting the theoretical model (11.30) to the experimental values gives an estimate of the practical range of the covariance.

11.5 Mean-Model Definition

In the previous section, we described a methodology for computing the bias correction that could be used with the "mean model" for robust optimization. We did not, however, discuss how to define and calculate the mean model. In numerical subsurface models, properties such as permeability appear as spatially varying coefficients in the partial differential equations describing fluid flow and are inputs to the numerical reservoir flow simulator. It might then seem sensible to define the mean model as the arithmetic mean of the ensemble of permeability fields. In ensemble-based data assimilation, however, updating the log-permeability fields and using the mean logarithm of permeability to estimate the gradient is standard. In this case, it might seem that the sensible decision is to use the geometric mean of the ensemble of permeability fields or, equivalently, the arithmetic mean of the log-permeability fields. Additionally, we note that the harmonic mean is useful when computing the effective permeability of beds in series.

The Box-Cox transformations (Box and Cox 1964) provide a continuum of transformations applicable to a property field before computing the mean. We define a variable with power λ as

$$\theta_\lambda(x) = \begin{cases} (x^\lambda - 1)/\lambda & \text{if } \lambda \neq 0 \\ \log(x) & \text{if } \lambda = 0, \end{cases} \tag{11.31}$$

and the mean of the transformed variables are computed in the normal way,

$$\overline{\theta_\lambda(x)} = \frac{1}{N}\sum_{j=1}^{N} \theta_{\lambda,j}(x). \tag{11.32}$$

The λ-mean of the original variable is then

$$\overline{x_\lambda(\theta)} = \begin{cases} (\lambda\,\overline{\theta_\lambda(x)} + 1)^{1/\lambda} & \text{if } \lambda \neq 0 \\ \exp\left(\overline{\theta_\lambda(x)}\right) & \text{if } \lambda = 0. \end{cases} \tag{11.33}$$

The arithmetic mean of the property field corresponds to using $\lambda = 1$, while $\lambda = 0$ and $\lambda = -1$ results in the geometric and harmonic mean definitions. In Sect. 11.7, we will discuss the consequences of applying various transformations before computing the mean model on the bias magnitude in a porous media flow problem. Note that while there may be an "optimal" definition of the mean model for most problems, using the optimally selected mean model is not necessary for the success of the minimization when using bias correction. A good choice of mean model (or, equivalently, a good choice for the parameter transformation) can make the bias correction easier to estimate with a small number of function evaluations or even make bias correction unnecessary.

11.6 Derivative-Free Optimization with Bias Correction

Suppose the objective function evaluated at the mean model with or without bias correction provides a suitable approximation of the robust objective function. In that case, we can minimize that single objective function. The options available for minimization are broad, and one is not necessarily limited to EnOpt-type algorithms, as described in Chap. 10.

Larson et al. (2019) provide a recent survey of derivative-free optimization algorithms, including a thorough review of model-based methods whose updates are based primarily on the predictions of a model that approximates the objective function. No one algorithm is best for all types of optimization problems. Still, Moré and Wild (2009) ranked model-based trust-region methods highest overall when investigating the performance of various derivative-free algorithms on a series of benchmark optimization problems with varying smoothness characteristics and dimensions. They evaluated the algorithms, not simply on the accuracy of the solutions, but on the fraction of the optimization problems they could solve to specified accuracy within a given computational budget. The highest ranking method was a function-interpolating trust-region method, NEWUOA, by Powell (2006), that fits a quadratic model of the objective function to $2N_u + 1$ evaluation points (N_u is the number of control variables). Note that a full quadratic model requires $(N_u + 1)(N_u + 2)/2$ coefficients, so the quadratic model in NEWUOA is highly under-determined for large N_u. Each successful iteration adds a better solution to the interpolation set based on the minimization of the quadratic model and drops the "worst" solution.

Algorithm 1 Model-based derivative-free optimization of mean model

Set $k = 0$
Initialize best-guess $\mathbf{u}_0$ and $2N_u$ evaluation points in $\mathbf{U}_k \in \Re^{N_u \times 2N_u+1}$.
Evaluate $\Omega(\mathbf{u}, \overline{\boldsymbol{\theta}})$ for $\mathbf{u} \in \mathbf{U}_k$ ▷ $(2N_u + 1)$ function evaluations
repeat
  Build a quadratic model $q_k(\mathbf{u})$ that interpolates $\Omega(\mathbf{u}, \overline{\boldsymbol{\theta}})$ for $\mathbf{u} \in \mathbf{U}_k$.
  $\mathbf{s}_k = \arg\min_{\mathbf{s}:\|\mathbf{s}\| \leq \delta_k} q_k(\mathbf{u}_k + \mathbf{s})$ ▷ Minimize quadratic model
  $\mathbf{u}_{k+1} = \mathbf{u}_k + \mathbf{s}_k$ ▷ New best point
  $\Omega(\mathbf{u}_{k+1}, \overline{\boldsymbol{\theta}})$ ▷ One function evaluation
  $\rho_k \leftarrow \frac{\Omega(\mathbf{u}_k, \overline{\boldsymbol{\theta}}) - \Omega(\mathbf{u}_k, \overline{\boldsymbol{\theta}})}{q_k(\mathbf{u}_k) - q_k(\mathbf{u}_{k+1})}$ ▷ Function/model ratio improvement
  if [ρ_k acceptable] **then**
    Augment $\mathbf{U}_k$ with $\mathbf{u}_k + \mathbf{s}_k$.
    Remove the point of $\mathbf{U}_k$ furthest from $\mathbf{u}_{k+1}$.
  end if
  $k \leftarrow k + 1$
until converged

One of the primary benefits of the function-interpolating trust-region methods is the efficiency gained by reusing sample points from previous iterations (Berahas et al. 2022). Finally, function-interpolating trust-region methods tend to be more robust in the presence of noise in the objective function than other techniques for derivative-free optimization and much more robust than gradient-based methods (Berahas et al. 2019). Although the number of function evaluations (reservoir simulator runs) scales linearly with the number of variables optimized in NEWUOA, the complexity of the quadratic model fitting may be the limiting factor in its applications to large problems as the number of operations required for fitting the quadratic model in even an efficient implementation scales as $\mathcal{O}(n^4)$ (Conn et al. 2009).

Applying a function-interpolating trust-region method to the problem of minimization of the mean model is straightforward. Freely available software codes are available for solving both the unconstrained optimization problem (Powell 2006; Ragonneau and Zhang 2024) and the bound constraint optimization (BOBYQA) problem (Powell 2009; Cartis et al. 2019).

Algorithm 1 provides a simplified summary of a function-interpolating trust-region derivative-free optimization algorithm to minimize the objective function evaluated on a single model. Note that although the algorithm requires $2N_u + 1$ function evaluations in the initialization step, subsequent iterations require a single function evaluation.[1]

[1] Bangerth et al. (2006) dismissed the NEWUOA algorithm as being too costly, stating that it requires five function evaluations at each iteration for the one-well 2D location problem. Correctly, it requires five function evaluations at the initialization step, then one function evaluation at all subsequent iterations.

If optimization using the mean model without bias correction is insufficiently accurate, Algorithm 1 needs to be modified so that the optimized function is the sum of the objective evaluated at the mean model and the bias correction. A continuous updating of the bias correction at each iteration is feasible but leads to relatively extensive modifications of the minimization algorithm. A more efficient approach is to perform first the minimization using the mean model without bias correction. Then, estimate the bias correction in the neighborhood of the mean-model optimization point and re-initialize the optimization using the bias correction.

We will present two applications using the mean model with bias correction to illustrate the benefits and limitations of the approach. In the first example, we investigate the reasonableness of the bias correction model and the impact of the choice of the mean model. The second case uses derivative-free optimization methods to compute the minimizer of the sample average approximation of the objective function's expectation.

11.7 1D Single-Phase Porous Flow Example

In this example, the problem is to optimize the location of a producing well in a one-dimensional reservoir, see Fig. 11.1, such that the magnitudes of the expected flow from each direction are equal. This objective is equivalent to minimizing the total flow into the producing well. We apply a constant pressure boundary condition at $x = 0$ and $x = 150$ and fix the pressure at the producer. The single-phase flow follows Darcy's law. Fluid viscosity and reservoir cross-sectional area are uniform and assumed to be known.

The permeabilities in this toy problem are uncertain, with a prior distribution for permeability log-normal with an exponential covariance, a practical range of 40, and a standard deviation of 2. We observe the log-permeability at the four locations shown in the left panel of Fig. 11.2, and the posterior uncertainty, conditioned on the observations, is characterized by an ensemble of draws from the posterior distribution, right panel of Fig. 11.2. The mean log permeability on the left side of the domain is larger than the mean log permeability on the right side because of the conditioning to observations.

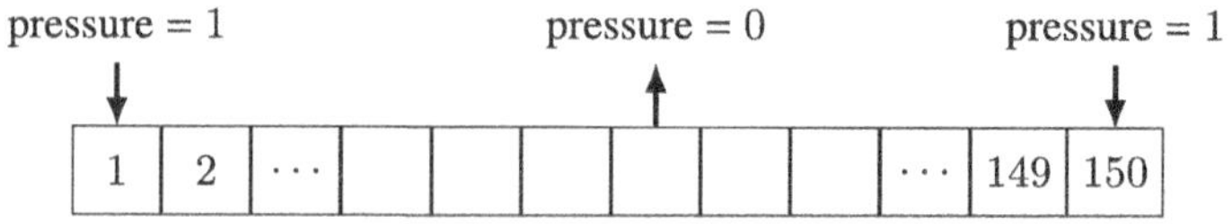

Fig. 11.1 One-dimensional model domain with 150 grid cells and pressure boundary conditions used to optimize the location of the producing well such that the expected production rate is maximized. Permeabilities in the grid cells are uncertain. (Reproduced from Fig. 1 in Oliver 2024)

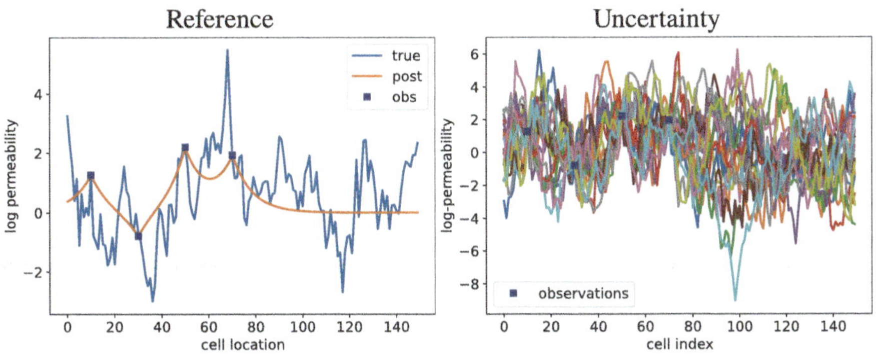

Fig. 11.2 The left plot shows the truth reference solution (blue line) and four log-permeability measurements used in an objective analysis to compute the posterior mean permeability estimate (orange line). The right plot shows an ensemble of simulated permeability curves that illustrate the prior permeability uncertainty, which we will use in the well-location optimization. (Reproduced from Fig. 2 in Oliver 2024)

The objective function for the 1D well location problem is inexpensive enough that we can use it to investigate the effect of the choice of property transformation on the magnitude of the bias when using the mean model for optimization. As discussed in Sect. 11.5, permeability and log-permeability are reasonable choices for the computation of a mean model to optimize the well location. Figure 11.3 (left plot) compares the objective function for the arithmetic mean of permeability (blue) to the sample average approximation of the expectation of the objective function (black). Two aspects of the comparison are fundamental. The first is that the difference in the two curves is quite significant—the sample average approximation of the expectation is around 2 for most of the reasonable range of well locations. At the same time, the objective function for the arithmetic mean of permeability is in the range 15–18. More importantly, the optimum well-location computed using the arithmetic permeability mean as the mean model is close to $u = 38$ (blue dashed line). In contrast, the correct optimum location is close to $u = 85$ (black dashed line).

The curves in Fig. 11.3 (right plot) show the objective function evaluated at the mean of a range of Box-Cox transformed permeability fields. A "mean model" with $\lambda = -0.5$ results in a tiny bias correction factor over a moderately extensive range of well locations. On the other hand, using the mean of log-permeability ($\lambda = 0.0$) results in a more extensive bias correction than obtained with $\lambda = -0.5$, but a better approximation of the shape of the objective function and a better estimate of the minimizer, even if we do not use bias correction. Optimization using the mean model defined with $\lambda = 0.2$ provides an excellent approximation of the "true" optimizer for the robust problem, even without bias correction. Note that while there may be an "optimal" definition of the mean model for most problems, using the optimal mean model is not necessary for the success of the optimization in those cases for which we are willing to apply the bias correction.

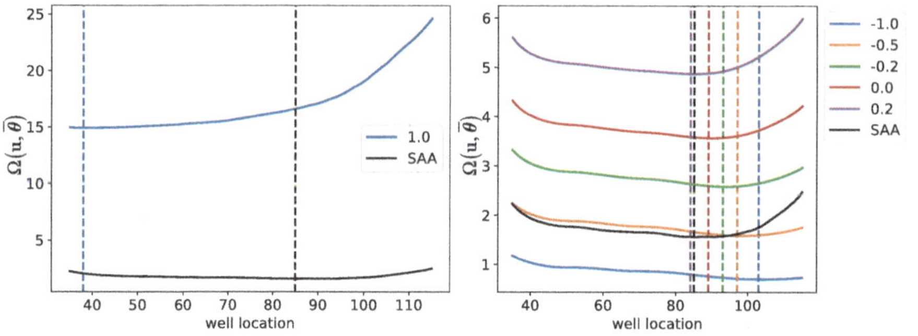

Fig. 11.3 Effect of type of averaging on the objective function for the "mean model." The left plot compares the objective function for the arithmetic permeability mean (blue) to the sample average approximation of the expectation of the objective function (black). The right plot shows the objective function evaluated at the mean of a range of Box-Cox transformed permeability fields together with the optimized solution for each case. (Reproduced from Fig. 5 in Oliver 2024)

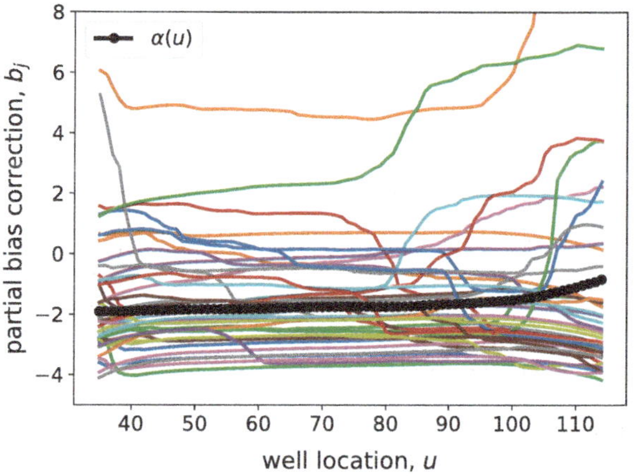

Fig. 11.4 The plot shows the partial additive bias corrections, $\beta(u)$, computed for 40 realizations of the model. (Reproduced from Fig. 6 in Oliver 2024)

Before leaving the 1D well-location optimization problem, it is valuable to briefly evaluate the appropriateness of the hierarchical model for bias correction. Figure 11.4 shows the bias correction function $\alpha(x^i)$ and 40 realizations of the b_j evaluated at all x. Note that the bias correction α (black dots) is smoother than most realizations of the partial bias correction, so it is potentially easier to estimate. We also note that we could perhaps model the mean partial correction (for fixed model $\boldsymbol{\theta}_j$) as Gaussian as in Eq. (11.11), i.e., each of the curves in Fig. 11.4 appears to have a characteristic "level" and fluctuations about that level, at least in the region of plausible solutions.

11.8 2D Two-Phase Porous Flow Example with Facies Model

The second numerical example also presents a problem of optimal well placement. Now, the domain is two-dimensional, the flow is two-phase, and the uncertain permeability field is piecewise constant—a simplified model of geologic facies. The upper plot of Fig. 11.5 shows the location of three producers near the model's southwest, northwest, and northeast corners. We wish to determine the location of an injection well that maximizes the reservoir production's expected net present value (NPV). We approximate the expected value by the sample average of 100 objective functions evaluated on realizations of the permeability field. The middle plot of Fig. 11.5 shows the first 12 permeability realizations.

The mean model in this example shown in the bottom plot of Fig. 11.5 is not perfectly uniform because we have approximated it from a finite 100 realizations ensemble of the log-permeability field. Despite the small variability, the mean model is much more uniform than any of the realizations, making optimizing the well location in the mean model easier than in the individual realizations. Unlike in the 1D well-location problem, the optimal well location in the 2D problem is sometimes on the boundary of the feasible domain; hence, for this problem, we use a bound optimization version (Powell 2009) of the minimization algorithm in the previous section.

Similar to the derivative-free optimization approach described in Algorithm 1, we initialize the minimization algorithm applied to the mean model without bias correction with the injector location at (1.25, 0.5) (Fig. 11.6). Because the number of control variables is 2 for the 2D well location problem, four additional locations spaced around the initial guess are included in the first evaluation set (Fig. 11.6). We needed nine function evaluations for convergence with the mean model when using "bound optimization by quadratic approximation" (BOBYQA) from Powell (2009).

The optimal injection well location depends strongly on the permeability field in this problem. Figure 11.7 shows the optimal well locations for each of the permeability realizations individually (blue dots) and the well position that optimizes the expected value of the objective function (red cross). The variability in the placement of nominal optimizers is large, but the optimizer of the mean log-permeability model (orange square) is relatively close to the robust optimizer. More importantly, the value of the expected objective function evaluated at the mean model, approximately equal to 78.7, is relatively close to the value obtained by robust optimization, close to 79.1.

As the estimate of the optimal location from the mean model is not identical to the robust optimal location, bias correction might be appropriate for this problem. Bias correction can, however, be expensive depending on how we compute it. In the 1D numerical flow example, we used a bias correction approach where we improved the bias correction at each iteration and each evaluation location. The benefit of the bias correction updates at later iterations was small in that approach because the magnitude of the bias correction was already well estimated in the neighborhood of

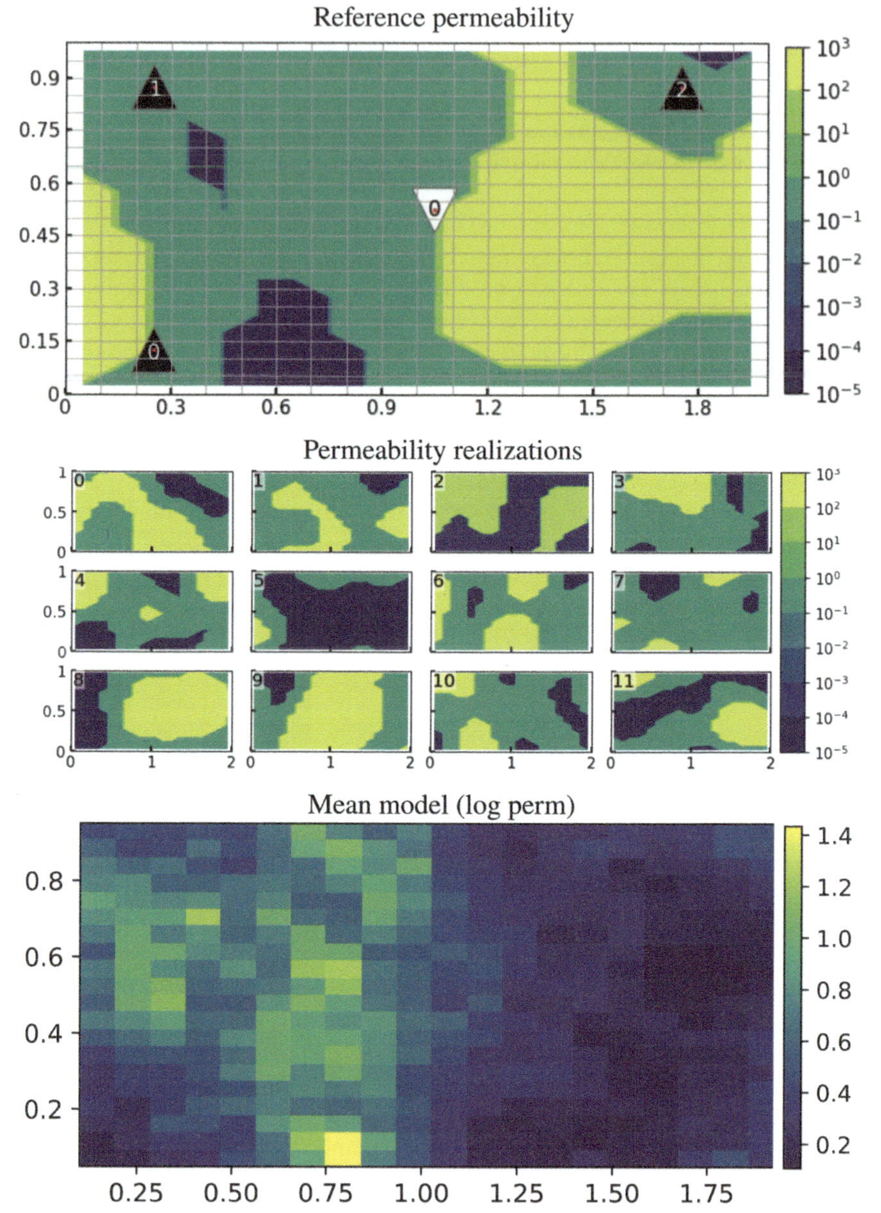

Fig. 11.5 The upper plot shows the facies field of the 2D reference model. The middle plot displays the first 12 facies realizations of the prior ensemble, while the bottom plot illustrates the permeability realizations' arithmetic mean model. (Reproduced from Figs. 7 and 8 in Oliver 2024)

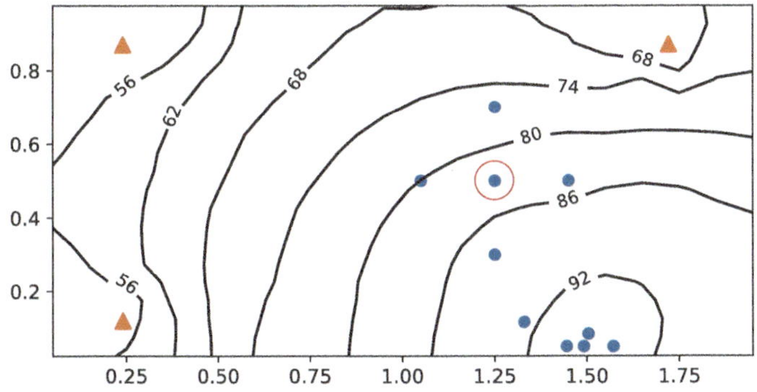

Fig. 11.6 Illustration of using the derivative-free optimization method BOBYQA to optimize injector location for the mean model without bias correction. The red circle indicates the injector's initial location. (Reproduced from Fig. 9 in Oliver 2024)

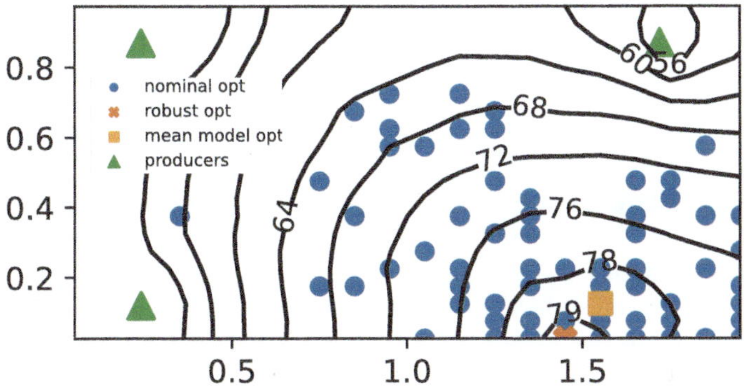

Fig. 11.7 Optimal well locations for nominal optimization, mean-model optimization, and optimization of the sample average of the objective functions for all realizations. (Reproduced from Fig. 10 in Oliver 2024)

the optimizer. Here, we take an alternative approach where we first estimate the mean model's optimizer without bias correction. Minimization without bias correction is generally an inexpensive operation, requiring only nine function evaluations for this problem. We then re-initialize the minimization by evaluating p_m model realizations at each mean-model evaluation location. To accurately estimate the hyperparameters of the hierarchical model for bias correction, we use the same model realizations at each evaluation location. We do not perform any further sampling after the initial estimation of bias correction.

Figure 11.8 shows the estimates of the bias correction for the situation where we perform the sampling of realizations at the initial evaluation locations only (blue dots). We have varied the number of function evaluations at each mean-model evaluation location using 10, 40, and 100 model realizations simulated at the locations

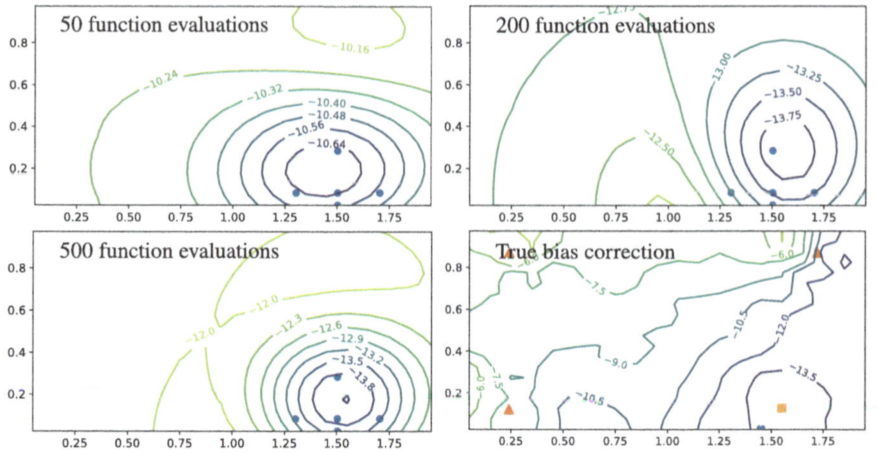

Fig. 11.8 Estimates of the bias correction obtained from multiple function evaluations at re-initialization locations. We use the same model realizations at each evaluation point. The numbers in each plot refer to the total number of function evaluations used for bias correction. If there are five locations and we use ten model realizations, there are 50 function evaluations. (Reproduced from Fig. 11 in Oliver 2024)

shown by the blue dots. The shapes of the three bias correction surfaces are generally similar. However, the magnitude of the estimated correction is significantly smaller for $p_m = 10$ than for $p_m = 40$ or $p_m = 100$. For comparison, we also computed the actual bias correction surface shown in Fig. 11.8 using 40,400 flow simulations.

In general terms, we identified three approaches to maximizing the expectation of the objective function in Eq. (11.4): one can seek to maximize the expectation directly, one can obtain an approximate solution using only the mean model in the optimization, or one can use the mean model with bias correction. When the objective function is nearly linear in the model parameters, optimizing the objective applied to the mean model should provide a good solution.

The dependence of the objective function (NPV) on the permeability field is, however, highly nonlinear in this 2D flow example, as evidenced by the large variability in the optimal location of injectors for individual realizations of the permeability field illustrated by Fig. 11.7. Somewhat surprisingly, perhaps, the estimations of an optimal well location for the expected value of the objective function using three different approaches give remarkably similar results, as seen in Fig. 11.9, even when the averaging method used for computing the mean model is not optimized. Figure 11.9 shows contours of the expectation of the objective function (unlike Fig. 11.7, which showed contours of the objective function evaluated on the mean model). Global optimization of Eq. (11.4) using exhaustive search required 40,000 function evaluations, obtaining the location indicated by the blue cross in Fig. 11.9. Using a mean model, computed using the ensemble of the log-permeability fields and the BOBYQA algorithm, gives a solution (orange square) with an objective that is quite close to the global optimum,

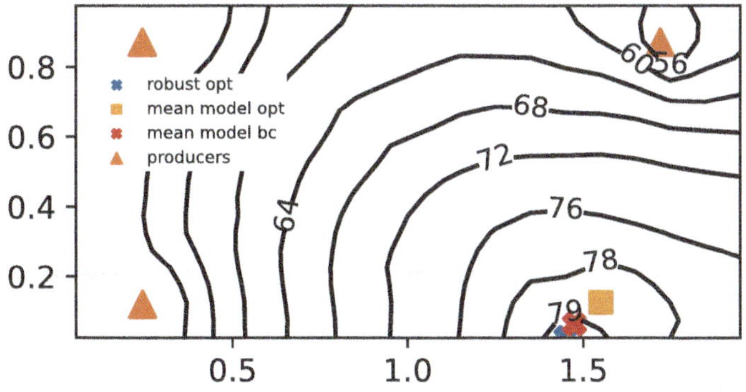

Fig. 11.9 Optimal well locations for nominal optimization, mean-model optimization, and optimization of the sample average of the objective functions for all realizations. (Reproduced from Fig. 12 in Oliver 2024)

but at the cost of only 10 function evaluations. Finally, restarting the minimization algorithm at the solution obtained using the mean model but adding bias correction results in a solution almost identical to the true solution (see the upper right plot in Fig. 11.8). The bias correction approach costs 10 function evaluations for the mean-model minimization without bias correction plus 200 function evaluations to compute the bias correction surface, plus six final function evaluations to minimize the bias-corrected mean-model objective function.

11.9 Summary

Optimization on the mean reservoir model has sometimes been used as an approximate method for reservoir management optimization under uncertainty because of its relatively low cost relative to the optimization of an expected value. However, we cannot generally know the appropriateness of that approximation without extensive computation. In this paper, we have suggested an additive bias correction method that allows for using a mean model with confidence for robust optimization. The cost of estimating the bias correction in our examples is considerably less than the cost of a sample average approach to evaluating the expectation of the objective function. The efficiency of the bias correction approach is enhanced when using a derivative-free optimization method that accounts for the sequential updating of the bias correction estimate.

We have demonstrated the method for two optimal well location problems. In the first problem, we investigated applying sequential bias correction in a derivative-free optimization method and examined the appropriateness of a hierarchical bias correction model. The first test problem was also to explore the effect of the choice of definition of the mean model on the need for bias correction. For some choices of the "mean," the bias in the estimated optimal control would be huge, and bias correction would be necessary. For other mean-model choices, the bias was so tiny that the solution obtained using the mean model without bias correction was the same as obtained from sample averaging of the objective function.

The second numerical example also examined the problem of estimating optimal well location when permeability is uncertain. This example, however, was more complex than the 1D well location problem; the objective function was the expected net present value of the reservoir, the flow was two-phase oil and water, the permeability field was piecewise continuous, and the reservoir was two-dimensional. We showed that we could implement a very efficient optimization method in which the mean model was first used without bias correction, followed by bias estimation in the neighborhood of the estimated mean-model optimizer, and a restart of the derivative-free optimization. Optimization using the mean model gave a well location whose expected NPV was only 0.2% lower than the optimal value at the cost of only six function evaluations. Adding even a tiny amount of bias correction gave the correct solution at the cost of 62 function evaluations.

Indeed, there will be many problems for which the mean of an uncertain quantity might not be meaningful, or the choice of a mean model might be difficult. If a "mean model" is chosen poorly in those cases, the bias correction will still provide a valid method for robust optimization but may not be as efficient.

Chapter 12
Closed Loop Reservoir Management

Abstract This chapter discusses closed-loop reservoir management (CLRM) workflows, which recursively and repeatedly combine ensemble history matching with robust optimization of decision alternatives, generating inputs to a decision process to determine the subsequent actions for the reservoir production or development strategies. Asset teams usually work in an annual maturation process (AMAP), using updated models to determine next year's strategies. The AMAP process fits well with the CLRM workflows discussed here.

12.1 Introduction

The ensemble-based closed-loop reservoir management (CLRM, see Jansen 2011) workflows, or more generally, closed-loop reservoir management and development (Mirzaei-Paiaman et al. 2021), connect ensemble history matching with robust optimization. After producing the reservoir through the current year, the history-matching method updates the reservoir description using the newly acquired production data. The data almost always contain new information about the reservoir that benefits the history matching and leads to reservoir models with better predictability. After that, based on the ensemble of updated model realizations, a robust optimization algorithm determines the "best" production and development strategies for the following year. The optimal strategy is input to a decision workflow that helps decide on the following year's operational plan. This repetitive annual maturation process (AMAP) is standard in many energy companies, and its connection to an ensemble-based CLRM workflow is state-of-the-art in operational reservoir management.

Chang and Evensen (2022) presented a CLRM example combining ensemble history matching and robust optimization to decide the optimal drilling order. The closed-loop decision workflow in Fig. 12.1 comprises various decision steps during field development with annual decision points regarding the drilling order of a sequence of production and injection wells. Chang and Evensen (2022) applied an iterative ensemble smoother together with ensemble optimization used recursively in time, although the workflow is independent of the methods used for history matching and optimization.

G. Evensen et al., *Ensemble History Matching*,
https://doi.org/10.1007/978-3-031-99155-4_12

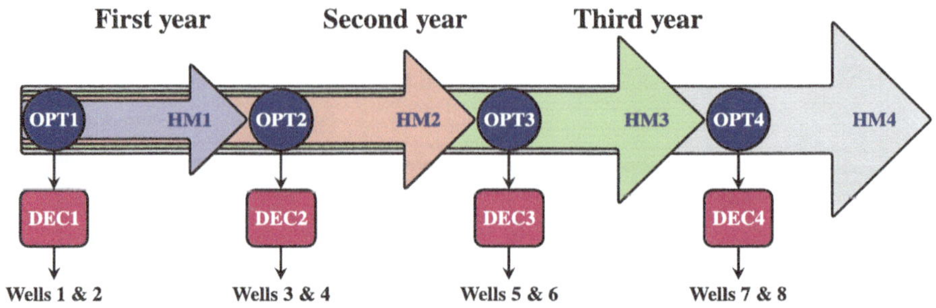

Fig. 12.1 Decision workflow for field development. Drilling two wells per year, we use the workflow to decide on the optimal drilling schedule and how many wells to drill (from Chang and Evensen 2022)

The workflow starts with the initial optimization OPT1, which leads to the decision DEC1, on which two wells out of eight planned ones provide the highest NPV when drilled first. After drilling the two wells, follow a year's reservoir production from the new wells while collecting the production rates. Then, the HM1 step conditions the reservoir model to all new and past observations to create the best possible ensemble of geologically consistent models fitting the measurements within their estimated uncertainty. The optimization step OPT2 finds the best alternatives, leading to the decision DEC2 for the subsequent two wells to drill. Chang and Evensen (2022) applied a simple static decision algorithm. They selected the two wells that gave the highest mean NPV averaged over the ensemble for the simulation over the next year. The decision to drill depends on obtaining a positive NPV. In Sect. 12.5, we will touch on more advanced dynamic decision methods that account for the impact on future decisions.

Because closed-loop reservoir management has two fundamental components, history matching and production optimization, and the history-matching task is more mature than the production optimization task, most research papers on closed-loop reservoir management have focused on the problem of production optimization under uncertainty.

Decision-making for optimal production or drilling schedule is straightforward, as we would choose the optimal controls or drill the optimized wells. However, making decisions based on an ensemble of predictions is more tricky. Assume we use our history-matched ensemble and simulate the future production from an optimized infill well over the whole ensemble. If the predicted ensemble of NPVs has a Gaussian distribution, we usually select to drill the well if the ensemble mean NPV is positive. However, we could introduce a human bias, being risk-taking or risk-averse, possibly dependent on the investment cost or the potential upside or benefit. It is also possible to simultaneously minimize the risk of poor outcomes during the optimization.

Unlike the situation for history matching, where we can demonstrate the usefulness in real-field cases, it is hard to verify the effectiveness of closed-loop reservoir management in an actual field because we can only perform the process once. It is not possible to compare results for one set of decisions with an alternative set of decisions. Consequently, we base all good examples of the value of closed-loop optimization on synthetic test cases. Using synthetic test cases, we can evaluate different strategies multiple times and demonstrate the value of closed-loop optimization. In this chapter, we describe lessons learned from the Brugge benchmark case described by Peters et al. (2010) (but see also the example from Chang and Evensen 2022) before we give an account of methods for robust decision-making, where we account for the reservoir uncertainty.

12.2 Lessons from Brugge

TNO in Holland designed the Brugge benchmark case (Peters et al. 2010) to evaluate methods for solving closed-loop optimization problems. The objective of the optimization was to maximize the net present value of the fictitious Brugge reservoir by controlling rates or pressures on 36 injectors and producers, given well logs, 10 years of production history, and an interpreted 4D seismic data set. The organizers provided the participants with 104 plausible realizations of the geology based on the well logs and three geologic scenarios to represent the uncertainty before assimilating production data.

At the end of the 10-year history-matching period, all wells became converted to smart wells, and each well then had three independently controllable completion intervals. The participants in the benchmark exercise sent their recommended controls for the next 20 years to TNO, who then applied the controls to the reference reservoir model, which was never released or shown to the participants. The participants presented their history matching and production optimization results at an SPE Applied Technology Workshop in Brugge, Belgium, in June 2008. Each participating group described the methodology that they used to solve the problem. TNO revealed the resulting NPV for each of the teams at the workshop.

First, we remind the reader that closed-loop optimization consists of two parts: history matching with uncertainty quantification and production optimization. Achieving good results requires that both parts perform well. Evidence for the importance of both aspects can be seen in the strong correlation between the quality of the history match to data and the value of the expected NPV achieved in the optimization illustrated by Fig. 12.2. Better history matching led to better reservoir management. The relationship was likely causative (i.e., the high NPV values resulted from the improved reservoir models). However, it is also possible that teams that were better at history matching were also better at optimization.

Groups that used many parameters to describe uncertainty achieved better history matches and NPV than those that only used permeability or porosity multipliers for reservoir model layers. Most teams applying ensemble Kalman filter-like methodologies used the ensemble of geological realizations that TNO had provided. However, one team regenerated the ensemble with all realizations generated from a single scenario. Additionally, one team added uncertainty in the relative permeability curves,

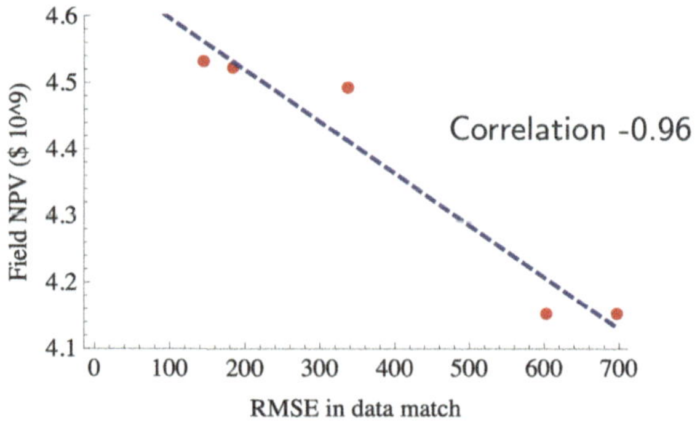

Fig. 12.2 In the Brugge study, results from the second round showed that better history matching resulted in better reservoir management

the vertical transmissibility, and the original oil-water contact to account for model error introduced by coarsening from the data-generating model.

Unsurprisingly, one could generate superior matches to data by updating a high-dimensional parameter state vector. Still, the fact that the models with high-dimensional parameter state vectors gave improved predictability might be unexpected. While the danger of overfitting is always present, the Bayesian approach with realistic priors on variables limited the risk of overfitting. Additionally, a parsimonious model designed only to match data in the first 10-year history period would not have provided sufficient degrees of freedom for the characterization of uncertainty in reservoir behavior after the introduction of smart completions. When expanding the control options at year 10 to allow controls of individual well completions, the uncertainty in forecasts increased substantially (see Fig. 12.3). This behavior would not have been possible with a parsimonious parameterization.

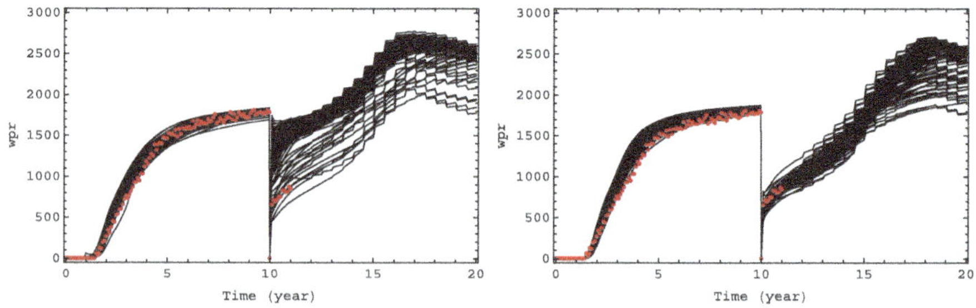

Fig. 12.3 The left plot shows the history match to the end of year 10 and predictions into the future, while the right plot shows the history match to the end of year 11. The introduction of independent completion controls after 10 years resulted in a much larger prediction spread. Red dots show water production rates from the true data-generating model and ensemble of forecasts (black)

Chen and Oliver (2010a) obtained the best results using an iterative ensemble-based history-matching approach (Gu and Oliver 2007) with localization of parameter updates and an ensemble-based optimization method (EnOpt) described in Chap. 10. The second-best results (Chen et al. 2010) were again obtained using ensemble-based history matching but applying the mean of the history-matched models with an adjoint-based minimization method for optimization. While this approach did not rigorously capture the effect of the uncertainty in the model because there was no correction for the bias (see Chap. 11), it appears to have provided a reasonable approximation of the objective function's expectation, and the adjoint system provided accurate gradients for minimization.

Although the Brugge benchmark case served as an example of closed-loop optimization, the participants had limited opportunities to revise their controls based on new data. In round two of the exercise, production data was provided for each group for years 10–20 based on their estimated controls from round one. The teams then re-history-matched the reservoir models and updated the controls. The only exception was the team of Chen and Oliver, who were allowed to update controls yearly after year 10 for the lifespan of the reservoir (Chen and Oliver 2010a). This approach was closer to the process typically followed in applying closed-loop optimization. The NPV achieved through this closed-loop process was nearly identical to the NPV achieved by the workshop organizers when the production optimization was performed on the reference reservoir model, indicating that a closed-loop process of repeated history matching and production optimization can potentially give nearly optimal results.

Since the participants at the workshop were only allowed to submit two sets of controls, there was minimal opportunity to apply controls whose purpose was explicitly to learn about reservoir behavior and then apply the new knowledge to improve the objective function. The only team allowed the opportunity to repeatedly history match and update controls (Peters et al. 2010; Chen and Oliver 2010a) did not explicitly pursue that option. Instead, Chen and Oliver (2010a) pursued a "naïve optimization," focusing only on the contribution of the controls to the direct increase of the objective. Despite that limitation, the results of Chen and Oliver (2010a) were nearly optimal, indicating that serendipitous learning may be sufficient in some cases. This learning is fortuitous, as the additional complication of including the value of information in the optimization would make most real reservoir management problems infeasible unless it were possible to identify a few key reservoir features that needed estimation (Wang and Oliver 2021).

Using a minimization method instead of manual optimization always led to the best results for the Brugge benchmark. An adjoint-based gradient descent method applied to the mean model was almost as good as the EnOpt approach applied to the ensemble of model realizations. The uncertainty was computed for highly parameterized models using ensemble-based history-matching methods in all top approaches. Methods that used small numbers of parameters for history matching were generally poor at history matching and reservoir management optimization.

The benefits of repeated history matching and production optimization were considerable, especially when there were significant control changes. For the Brugge case, the results obtained using closed-loop optimization were as good as those obtained with perfect knowledge of the reservoir.

12.3 Robust Decision-Making

The complexity of decision-making (Howard and Abbas 2015) and optimization in field development planning and reservoir management has increased significantly during the last few years (Bratvold et al. 2009). The transition from large, individual exploration projects to near-field exploration represents an essential shift. As a result, the decision-making problem, which previously focussed on isolated assets, now considers area development strategies. In these cases, fields and assets are often at different stages of their life cycles. Some are in the early phases of development, while others may be in mature production or nearing abandonment. The static geological and dynamic reservoir simulation models can vary widely regarding modeling complexity, data quality, and information availability. In addition, production and development strategies vary between assets, influenced by their specific challenges, goals, and constraints.

The demand to reduce oil and gas operations' environmental and climate impact has required minimizing greenhouse gas emissions and optimizing energy usage. This demand has significantly complicated the optimization process, and there is now a need to balance the maximization of economic performance with the minimization of the environmental impacts from CO2 emissions, flaring, and water management. At the same time, these decisions must ensure operational feasibility and safety (Aghajani et al. 2025; Zou et al. 2016).

Effective decision-making in reservoir management requires a systematic approach to handle the inherent uncertainties represented by the ensemble of reservoir models, i.e., what we denote robust decision-making. We will next explore decision-making frameworks suitable for reservoir engineering, emphasizing the transition from traditional static methodologies to more dynamic, adaptive strategies.

12.4 Static Decision-Making and Its Limitations

The history-matching output is an updated posterior ensemble of models consistent with the prior geological concept. It has a reduced spread and, most importantly, a higher predictive power than our initial prior ensemble. The updated ensemble becomes an input to the decision-making process for reservoir management. In CLRM, this approach translates into a robust optimization problem where we must choose the objective function, the control or decision variables, and the constraints.

The robustness comes from introducing the uncertainty using an ensemble of geological models when evaluating a decision's impact.

For example, we have an ensemble of geological models at a particular time, possibly conditioned on the previous production history and geophysical data. Based on this ensemble, we can simulate different decision alternatives, e.g., regarding well placements or drilling orders, and evaluate each decision's ensemble mean NPV. Typically, we would choose the alternative with the highest mean NPV if it is greater than zero. We can also obtain decision alternatives from a robust optimization method. When we recompute the optimization recursively in time, it constitutes a closed-loop reservoir management workflow (Chang and Evensen 2022).

The robust optimization process should not be understood as a method to find the optimal strategy or the best control values but as a framework for decision maturation. The "optimal" strategies or control values the algorithm identifies often differ from their initial guesses. These differences come from utilizing an ensemble of models, which provides deeper insights into the reservoir's subsurface characterization and dynamic behavior. When the effects and implications of optimal strategies are well understood, we can either confirm hypotheses and solutions previously proposed by the engineers or obtain new insights and solutions.

The optimization step in the classical CLRM approach assumes that the data and level of uncertainty available at the current time remain static over the forecast horizon. While straightforward to implement, this approach often results in suboptimal long-term decisions. Although classical CLRM can incorporate new data into the repeated history-matching steps to update models and improve control optimization, it leads to a myopic strategy that fails to account for the possibility of intentionally increasing information by considering the value of information. Consequently, classical CLRM is increasingly unsuitable for addressing the complexities of modern reservoir management, especially early in a reservoir's life when the potential for reducing uncertainty through data is high.

The static optimization methods, also known as greedy or myopic optimization, focus on immediate rewards without considering the impact of future data gathering, learning, and decisions. These methods evaluate the immediate benefit of each decision at a given stage, selecting actions with the highest local payoff. The methods assume the decision-maker's understanding of underlying conditions, such as geological models, uncertainties, and market prices, remains time-invariant, and the decision-maker makes the decisions based on current knowledge without incorporating the potential for learning or adaptation. Static optimization methods are computationally efficient and easy to implement but only provide approximate solutions for time-dependent reservoir development scenarios, as they implicitly assume that current decisions will only weakly influence future reservoir development. The static methods' myopic nature ignores the decisions' long-term impacts. Additionally, they do not provide dynamic feedback since they do not adapt to updated data or uncertainties. Hence, the static methods are less suitable for decision-making in complex, time-dependent reservoir development projects.

12.5 Dynamic Optimization Methods

Dynamic optimization methods, particularly stochastic dynamic programming (SDP) (Howard 1960) and approximate stochastic dynamic programming (ASDP) (Powell 2011), improve decision quality by explicitly considering the evolution of uncertainties and decisions over time. Each decision is evaluated based on immediate rewards and its influence on future outcomes. Dynamic optimization offers several advantages, including generating optimal short- and long-term strategies, balancing multi-objective trade-offs, and robustly handling uncertainty through probabilistic models. However, SDP is computationally intensive and is often impractical for real-world applications due to the high dimensionality of state and action spaces, a problem known as the curse of dimensionality (Sutton and Barto 2018).

The development of ASDP techniques has addressed the challenges of implementing SDP models. These methods reduce the need for a complete model and lower computational costs by approximating value functions and policies. One of the most frequently used ASDP methods is the least squares Monte Carlo (LSM) approach (Longstaff and Schwartz 2015; Gamba 2003; Stentoft 2004). LSM simplifies dynamic optimization by approximating the value function, which represents the expected payoff of future decisions, using regression techniques applied to simulated data. Monte Carlo simulations generate an ensemble of possible future realizations, capturing uncertainties relevant to the decision. Using backward induction, LSM estimates the conditional expected value of future rewards, applying regression techniques—typically polynomial approximations—to map the value function as a function of state variables. This approach efficiently evaluates conditional expectations without explicitly enumerating all state-action pairs (Ahmadi and Bratvold 2023). The curse of dimensionality in SDP arises because computational and memory requirements grow exponentially with the number of state variables, action choices, and uncertainty dimensions. While LSM also grows exponentially with the number of state variables and action choices, it scales approximately linearly with the number of uncertainty dimensions (Powell 2011). This efficiency makes LSM highly applicable to real-world problems such as reservoir management, financial options pricing, and energy production planning, where decision spaces are large and uncertainties are significant. By balancing computational efficiency with accuracy, LSM delivers near-optimal policies for complex, high-dimensional decision-making contexts (Aghajani et al. 2025; Willigers et al. 2011; Thomas and Bratvold 2015).

12.6 Summary

In conclusion, static optimization methods, such as greedy or myopic approaches, are helpful for not very complex, time-sensitive problems with minimal future impacts of current decisions. However, they cannot incorporate future learning, which is essential in complex, uncertain, multi-stage decision-making contexts. Dynamic optimization methods, such as stochastic dynamic programming (SDP) and approximate stochastic dynamic programming (ASDP), address this limitation by accounting for evolving uncertainties and the value of future learning in current and future decisions. The least squares Monte Carlo (LSM) method is a robust and widely used ASDP technique that enhances computational efficiency. Using Monte Carlo simulations and regression, LSM efficiently evaluates decisions across multiple scenarios, making it a practical alternative to traditional SDP for high-dimensional problems. While ASDP methods, including LSM, are still computationally demanding, they mitigate the curse of dimensionality more effectively than SDP, making them feasible for real-world decision contexts such as reservoir management and renewable energy investments.

Ensemble decision methods are suitable for solving reservoir engineering problems since they allow for the simulation of multiple scenarios. These methods enable reservoir engineers to choose between complex trade-offs, optimize multi-objective goals, and respond to evolving conditions. Although the closed-loop approach remains valuable for straightforward reservoir management tasks, the demands of contemporary reservoir engineering—such as area development strategies and environmental considerations—necessitate more advanced methodologies. By adopting adaptive decision-making frameworks, reservoir engineers can improve decision outcomes, reduce risks, and achieve enhanced long-term reservoir performance.

Part III
History-Matching Examples and Analysis

In this part we discuss the properties of the history-matching problem by studying in detail a simple example. Thereafter we apply the approach to a large reservoir model and we demonstrate the use of ensemble methods in decision processes.

Chapter 13
History Matching the REEK Model

Abstract This chapter presents a simple but realistic history-matching case for a model of reservoir named REEK that we use to illustrate the crucial properties of the history-matching problem when conditioning on production rates. We start with a standard HM procedure computing global updates and demonstrating how we improve the results when properly accounting for uncertainties in model controls, dependency of measurement errors, and introducing localization schemes to reduce the impact of the spurious correlations appearing when using a limited sample size. We run multiple experiments and demonstrate correlation-based adaptive localization for the history-matching update. The history-matching problem is inherently under-determined because many different parameter updates give an equally good fit to the data. The final algorithm solves for the minimum geologically consistent update of the uncertain model parameters, providing a good fit to the data.

13.1 Introduction

The following examples are presented to provide an in-depth understanding of the properties of the ensemble history-matching problem in a practical setting. Several previous studies have pointed out issues with too low posterior uncertainty and even characterized the low uncertainty as an "ensemble collapse." The following discussion will demystify these issues and provide a basis for setting up a consistent history-matching experiment. We start in the following section by introducing the reservoir simulation model used in the experiments. After that, Sect. 13.3 discusses the formulation of the history-matching problem and the inclusion of uncertain controls in the state vector. In Sect. 13.4, we will see that the standard history-matching formulation with global analysis leads to excessive and violent model parameter updates and a near ensemble collapse. The remainder of the chapter will present several modifications to the history-matching workflow that will improve the results and consistency.

Section 13.5 examines the impact of ensemble size on the global update. An important conclusion is that when using up to one hundred realizations, the spurious correlations may reach similar amplitudes as the physical correlations. The predicted

G. Evensen et al., *Ensemble History Matching*,
https://doi.org/10.1007/978-3-031-99155-4_13

well rates relate nonlinearly to the reservoir parameters, and for this example, we found a maximum correlation of around 0.45 between rates and petrophysical fields. Thus, one conclusion is that we must use a larger ensemble size to have a significant sample-noise to physical correlation ratio.

Evensen and Eikrem (2018) showed that the assumption of independent errors in the time series of observed production rates is often inappropriate due to the negligence of time-correlated allocation errors and redundancy in the rate data. Thus, rather than conditioning on a high-resolution time series of redundant rate data, we can subsample the data significantly without deteriorating the results. Below, in Sect. 13.6, we demonstrate how a case where we condition only on the total accumulated oil, gas, and water production results in a weaker update and retains a good match with a more significant and consistent uncertainty.

Next, in Sect. 13.7, we will evaluate the use of correlation-based localization, which reduces the impact of spurious correlations and allows us to use a smaller ensemble size. We still need a large ensemble to differentiate between physical and spurious correlations.

In Sect. 13.8 we combine the use of localization when conditioning only on the total accumulated production rates. This case results in a minimal update of the porosity field localized around the production-well locations and an excellent match to the observed rates. *We hypothesize that the results obtained by conditioning on the total accumulated production will be similar to those obtained by conditioning on the full-time series of observed rates while accounting for measurement dependency by specifying a full measurement error covariance matrix.*

13.2 The REEK Reservoir Model

All experiments in this chapter will use a simplified model derived from a realistic Equinor reservoir case. Equinor's researchers have used versions of this model during the previous 10 years as a benchmark case for testing new technologies and algorithms for HM and optimization. The model has 27,755 active grid cells on a 40-times-64 grid with 14 layers. Simulating a single realization over the historical three-year period takes about four minutes of CPU time on a single core. Hence, the model is sufficiently small and, thereby, fast enough to run multiple ensemble simulations, and it is convenient for testing new methods and algorithms. We provide the model with monthly historical production and injection data. The simulator computes the corresponding reservoir volume from the production data and uses it to force the simulation. The model is the same as was used by Evensen et al. (2019), Leeuwenburgh et al. (2016), Hanea et al. (2017), and Wang and Oliver (2019), although with different configurations for the uncertain parameters. These papers considered problems ranging from examining history-matching methods to testing ensemble optimization and closed-loop reservoir management systems.

However, the current model configuration differs from the original model received from Equinor. In the following experiments, *the only uncertain parameter is the*

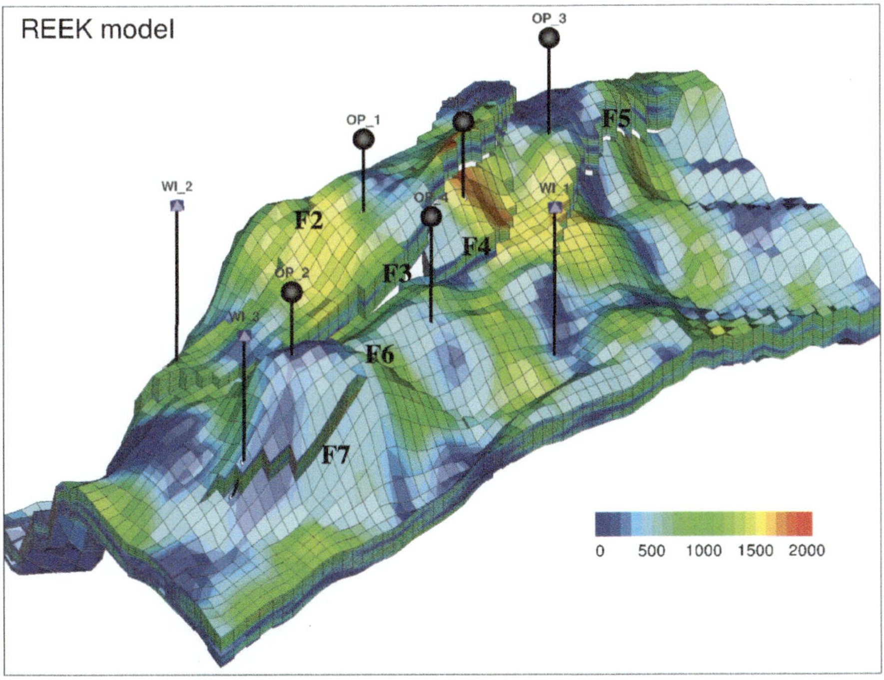

Fig. 13.1 Overview of the REEK model with well locations. This plot was generated using the Tecplot RS from Tecplot Inc

three-dimensional porosity field. Previous publications also considered uncertainty in the permeability field and uncertain fault multipliers. *We set the model's fault multipliers to zero, so all faults are closed, and the permeability field is the reference permeability defined for the model.* Thus, the history-matching problem is relatively simple, with only the porosity as an uncertain field variable, but this suffices for the discussion in this chapter.

There are five producing wells, OP1 to OP5, and three injectors, I1 to I3. See Fig. 13.1 for a three-dimensional representation of the model. We notice the strong faulting and how different producers drain different model compartments, partly separated by the faults. Injector I2 mainly supports OP1, injector I3 supports OP2, and injector I1 provides support for OP3, OP4, and OP5. *The historical production data conditioned on are the ones provided in the original reference model.* However, the porosity field ensemble consists of pseudo-random porosity realizations simulated using variograms computed from the reference model. *Thus, the ensemble of realizations does not contain the model used to generate the historical rates.* The REEK reservoir model is a synthetic case, and there could be no model error. However, we almost certainly have model errors in this case study because we only allow the porosity field to be uncertain and adjusted during history matching.

13.3 Defining the History Matching Problem

When integrating a reservoir simulation model over a past "historical period," we will generally force it by specifying a time series of the reservoir volume produced for each well. The model computes the reservoir volume rates from the historical time series of observed oil, gas, and water rates for each well. As far as possible, the simulation model will respect the imposed reservoir volume rate for each well while determining the individual production rates of oil, gas, and water production. Therefore, the historically produced reservoir volume or the rate observations act as control variables driving the simulation model. In contrast, the model parametrization and initialization will determine the relative proportions of produced oil, gas, and water volumes.

Evensen (2021) pointed out an inconsistency in the standard formulation of the HM problem. He suggested that we commonly include a measurement uncertainty on the historical rate data when we use these data in the conditioning. Still, we ignore the rate uncertainty when we use the historical rates to compute the reservoir volumes used to force the ensemble of reservoir simulations. By neglecting the uncertainty in these controls, we are not honoring all uncertainties, and we will underestimate the prediction uncertainty. As pointed out by Evensen (2021), including uncertainties in the historical rates leads to a significantly higher prediction uncertainty, particularly with rate uncertainties correlated in time.

We can write the reservoir model as

$$\mathbf{y} = \mathbf{g}(\mathbf{z}) = \mathbf{g}(\boldsymbol{\theta}, \mathbf{x}_0, \mathbf{u}), \tag{13.1}$$

where $\boldsymbol{\theta}$ includes all the uncertain static model parameters, $\mathbf{x}_0$ holds the model's initial conditions, and $\mathbf{u}$ contains the time-dependent uncertain controls. When we have uncertain controls, they also become part of the state vector $\mathbf{z}^\mathrm{T} = (\boldsymbol{\theta}^\mathrm{T}, \mathbf{x}_0^\mathrm{T}, \mathbf{u}^\mathrm{T}) \in \Re^n$, and consequently, we will also update the control ensemble together with the static model parameters in the analysis step. So, $\mathbf{u}$ is a vector containing the time series of historical rates for each production well from which we compute the produced reservoir volumes. We include the impact of the historical rate uncertainties by using an ensemble of control vectors. Introducing uncertainty to the historical rates leads to a more extensive spread in the ensemble predictions, which are otherwise only impacted by the geological uncertainty. This additional spread is most apparent in the ensemble of simulated rates early in the integration.

In the history matching, we condition the model on historical oil, gas, and water rates and update model parameters such that the relative fractions of produced oil, gas, and water are similar to the observed historical rates.

13.4 Introducing Control Uncertainty in the Standard HM Update

In this section, we will discuss what we denote "the standard reservoir case," characterized by using 100 realizations in the ensemble, ignoring control uncertainties, assuming uncorrelated measurement errors while conditioning on a time series of rate data, and finally computing a global update. In this case, we use the original historical rate observations to generate a single time series of produced reservoir volumes, which we use to force all model realizations. In addition to this standard case, we will present another case where we introduce the control uncertainty to establish a consistent formulation of the history-matching problem. See Table 13.1 for the case specifics.

In Fig. 13.2, we present in the left column the prior ensemble mean and standard deviations for the porosity field for this ESMDA100 case. The estimated ensemble mean and standard deviation from this "standard reservoir case" are plotted in the center plots. In contrast, the rightmost plots show the results from experiment

Table 13.1 The table shows a summary of experiment names and corresponding parameters. Here, "N" denotes the ensemble size, "Controls" indicate if the historical rates are uncertain and have been added to the state vector, the "Rates/Total" tells us if we are conditioning on the rate time series or the total accumulated production, "Global/Local" defines the assimilation scheme as using either global or local analysis, and in all cases, we use a diagonal measurement error covariance matrix. The local analysis is, in all cases, the adaptive one with default correlation truncation at $3/\sqrt{N}$ (see Chap. 7). In these experiments, we didn't use the measurement error inflation for the measurements with correlations close to the truncation value, as discussed in Sect. 7.5, even though we have indications that this procedure can significantly improve the results

Case name	N	Controls	Rates/Total	Global/local	C_{dd}
ESMDA100	100	No	Rates	Global	Diagonal
ESMDA100C	100	Yes	Rates	Global	Diagonal
ESMDA200C	200	Yes	Rates	Global	Diagonal
ESMDA400C	400	Yes	Rates	Global	Diagonal
ESMDA1000C	1000	Yes	Rates	Global	Diagonal
ESMDA100CT	100	Yes	Total	Global	Diagonal
ESMDA200CT	200	Yes	Total	Global	Diagonal
ESMDA400CT	400	Yes	Total	Global	Diagonal
ESMDA1000CT	1000	Yes	Total	Global	Diagonal
ESMDA100CL30	100	Yes	Rates	Local	Diagonal
ESMDA200CL21	200	Yes	Rates	Local	Diagonal
ESMDA400CL15	400	Yes	Rates	Local	Diagonal
ESMDA100CTL30	100	Yes	Total	Local	Diagonal
ESMDA200CTL21	200	Yes	Total	Local	Diagonal
ESMDA400CTL15	400	Yes	Total	Local	Diagonal

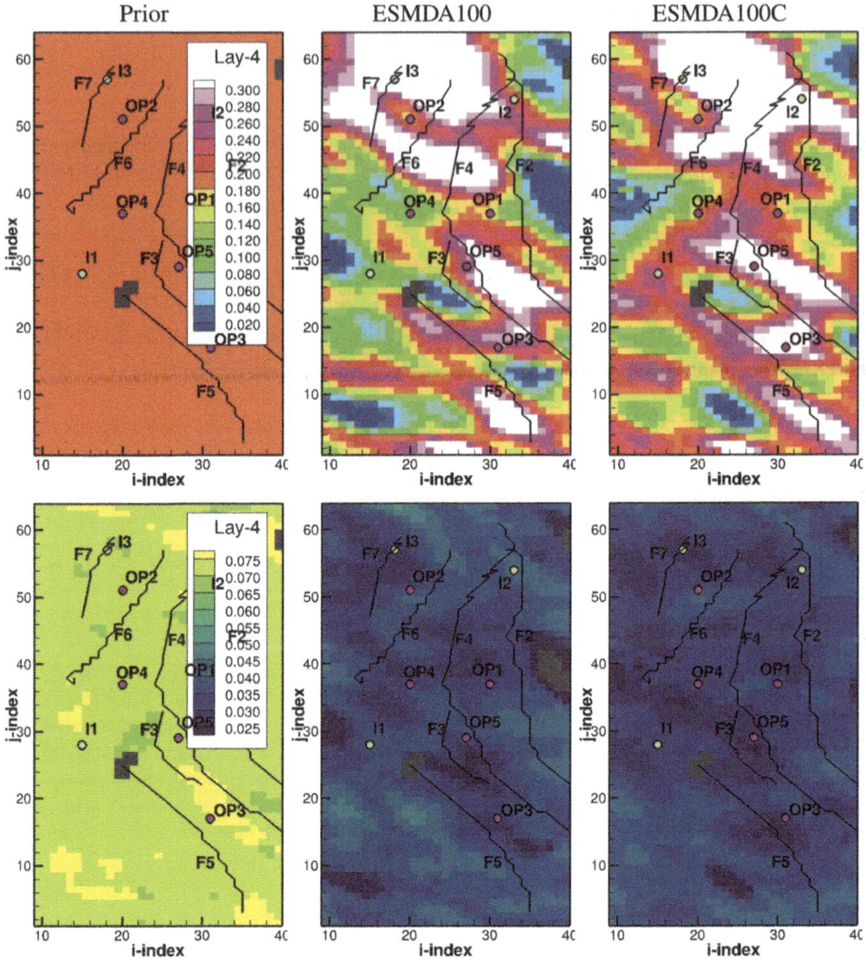

Fig. 13.2 The left column shows the prior porosity in layer four in the upper plot and its standard deviation in the lower plot. These plots represent the prior mean and uncertainty for all the experiments in this chapter. The second column presents the result of the standard reservoir case without any control uncertainty, and the third column gives the corresponding results from the case when we introduce uncertainty in the controls as discussed in Sect. 13.4. All contour plots in this chapter were generated using Tecplot 360 from Tecplot Inc

ESMDA100C where we have included the control uncertainty. The porosity updates from these two cases are relatively similar, and the impact of introducing the control uncertainty is not apparent here. The main things to notice from these plots are the excessive and violent porosity updates, which reach unphysical values at some locations and dominate throughout the model domain, also far from the well locations. Furthermore, the uncertainty is unrealistically small, almost zero, everywhere in the model domain. We have experienced something that we denote as "ensemble collapse."

In Fig. 13.3, we show the simulation results for one well from experiment ESMDA100. The left plots show the prior (green) and posterior (red) oil, gas, and water rates from top to bottom. In the right column of plots, we show the prior and posterior historic rates used to force the ensemble of model simulations. The color-shaded areas denote the ensemble mean plus and minus three standard deviations of the ensemble spread. In this case (ESMDA100), we note that the posterior ensemble prediction has an almost negligible uncertainty and does not sufficiently cover the historical rates. However, we can conclude that we match the data reasonably well. Note also that the ensemble spread at the initial time is zero since all wells will attempt to produce the same volume, and we have not perturbed the initial oil, gas, and water saturations. The uncertainty in the produced rates develops over time, given that the geological uncertainty impacts the simulations. Again, this case resembles the standard history-matching setup solved in most published works.

In Fig. 13.4, we show the corresponding plots for case ESMDA100C, where we have included uncertainty in the controls. We note that the prior initial uncertainty is now significant and more realistic as we force all model realizations by individual control realizations. Furthermore, we compute updated controls with reduced uncertainty, which we use to force the simulation of the posterior realizations. Including the uncertain controls has a minuscule impact on the update in this case. The posterior variance is still too low.

To conclude from these two cases, we are computing excessive and unphysical porosity updates throughout the model domain, and the posterior ensemble has collapsed to nearly zero variance, although the match to the historical data is acceptable. It is clear that the posterior porosity fields are unphysical, and the model will not have predictive capabilities, e.g., if we wish to simulate a new infill well. In all the following experiments, we will include the uncertain controls in the state vector and update them during the data assimilation.

13.5 Impact of Ensemble Size on the Global Update

One reason for underestimating posterior ensemble variance is related to sampling errors resulting from using a finite ensemble size. In many ensemble-HM applications for reservoir models, it is common to use an ensemble size of $N = 100$, a "magic" number typically inherited from ensemble data assimilation experiments in other

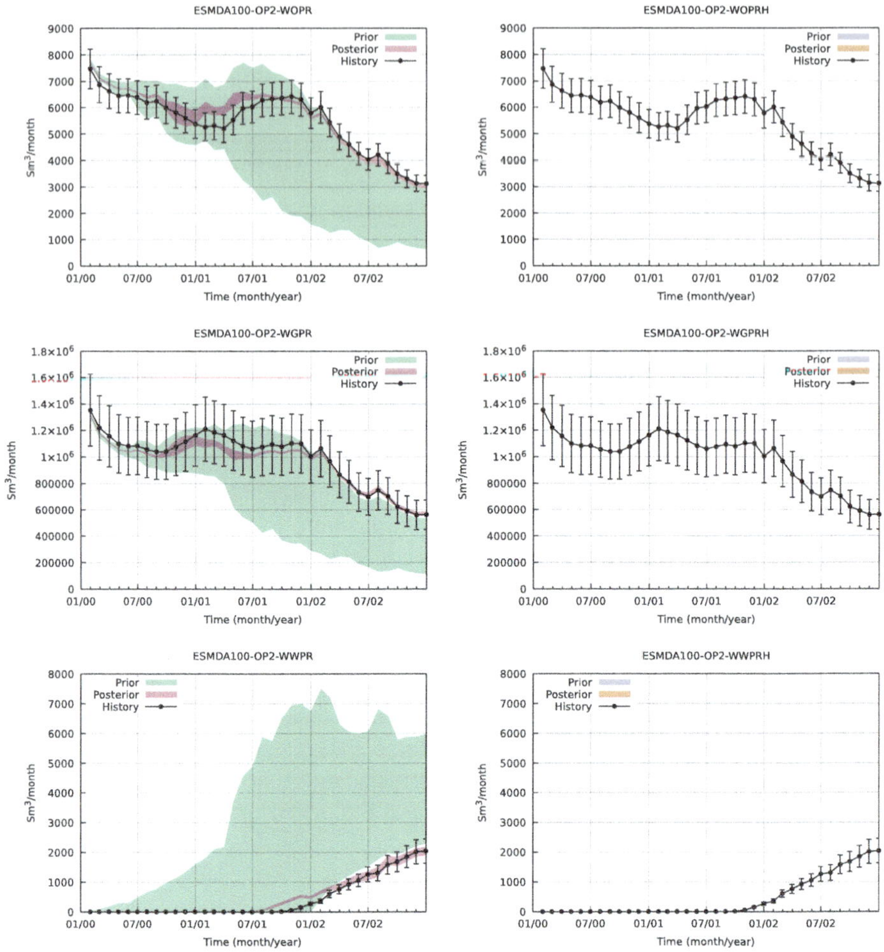

Fig. 13.3 ESMDA100 $N = 100$, global analysis conditioning on time series of rates using a diagonal $\mathbf{C}_{dd}$, without control uncertainty. In the left plots, we show the prior (green) and posterior (red) oil, gas, and water rates for OP2 from top to bottom. In the right plots, we show the prior and posterior historic rates used to force the ensemble of model simulations, which, in this case, without control uncertainty, become just the historical rates. All production profile plots in this chapter were generated using Gnuplot

fields such as oceanography and meteorology. Sampling errors decline proportionally to $1/\sqrt{N}$, and we will now examine the impact of increasing the ensemble size.

Figure 13.5 shows the updated ensemble mean and standard deviation for the global analysis with different ensemble sizes. It is clear that the larger the ensemble size, the more confined the updates are to the well locations and the lower the amplitudes of the updates. In the $N = 100$ case, we have implausibly large porosity updates with large amplitudes throughout the model domain. The strength of

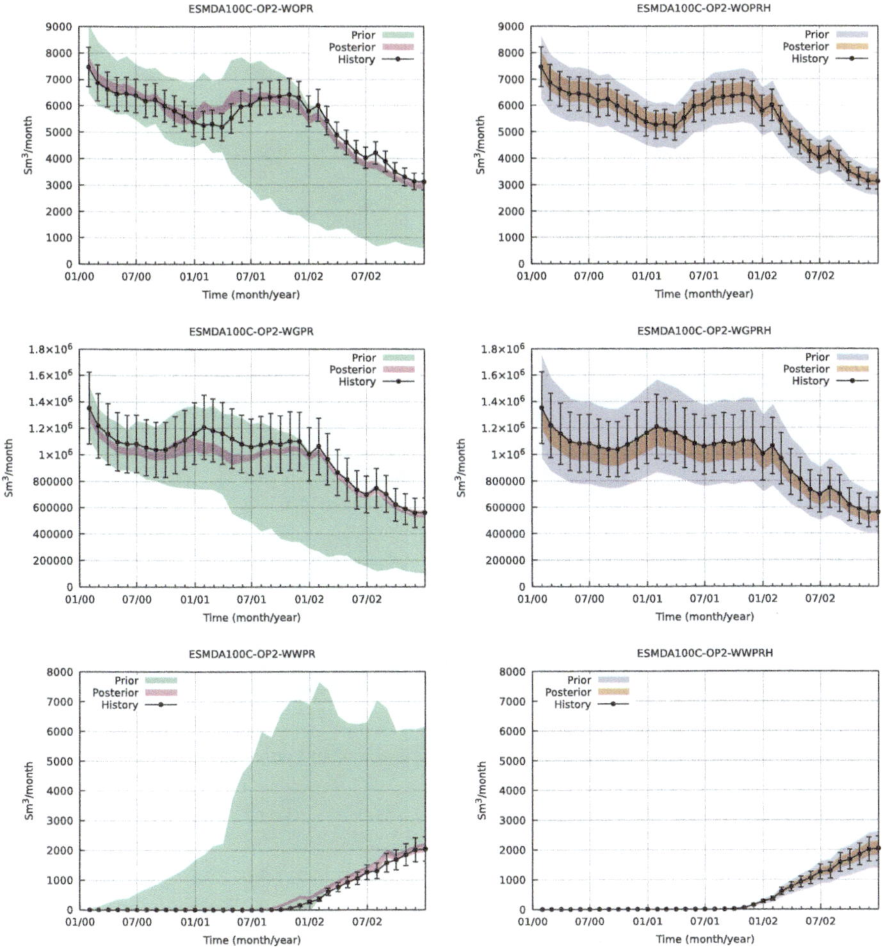

Fig. 13.4 ESMDA100C: $N = 100$, global analysis conditioning on time series of rates using a diagonal $\mathbf{C}_{dd}$ and including control uncertainty. Note the higher uncertainty at the initial time compared to case ESMDA100 and the finite uncertainty and updates for the controls in the right plots

the updates reduces significantly with increasing ensemble size. We also note that with growing ensemble size, we will have more significant variance in the posterior ensemble, and we will avoid the near "ensemble collapse" seen in the case of only 100 ensemble members. We also note that in the case of 1000 ensemble members, we have the most significant error reductions at the well locations, and we retain more of the variance further away from the wells.

The cause of the extensive updates and the substantial ensemble variance reduction far from the wells is connected to sampling errors often denoted as spurious correlations. We saw in Chap. 7 that the sample correlation between a measurement and any

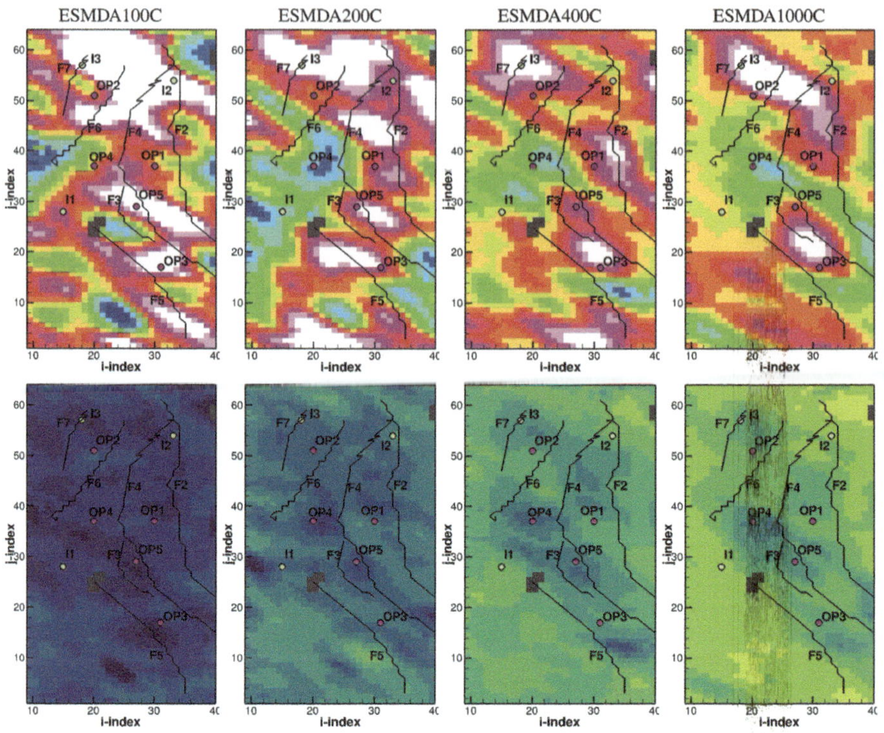

Fig. 13.5 The plots show the global layer four updates of the porosity mean (top) and standard deviation (bottom) using different ensemble sizes

state variable when the correct physical correlation is zero is a sample of a Gaussian distribution with a standard deviation equal to $1/\sqrt{N-3}$. We will ignore the subtraction of 3 in the following discussion, as we are concerned with relatively large ensemble sizes N. This theoretical result means that each observation will impact the update by its spurious sample correlation, and even if the spurious correlations from many observations tend to cancel each other out when updating the ensemble mean, they will all be interpreted as significant information and lead to a reduction of the ensemble variance. From Fig. 7.3, we find that, for a given ensemble size N, in 99.7% of the cases, the sample correlation will be in the range $(-3/\sqrt{N} : 3/\sqrt{N})$, and this result fits well with the plots in Figs. 13.6 and 13.7, when we compute the values of $3/\sqrt{N}$ for $N = 100$, $N = 200$, $N = 400$, and $N = 1000$, respectively as 0.3, 0.21, 0.15 and 0.095 (Figs. 13.8, 13.9 and 13.10).

We will next plot some correlation functions and see how they depend on the ensemble size. Figures 13.6 and 13.7 show, from left to right, the correlation functions between the oil, gas, and water production rates of well OP_1 at month 25 and the porosity field in the model's layer four. We chose layer four as this is the layer with the strongest correlations. From top to bottom, we plot the correlations using an

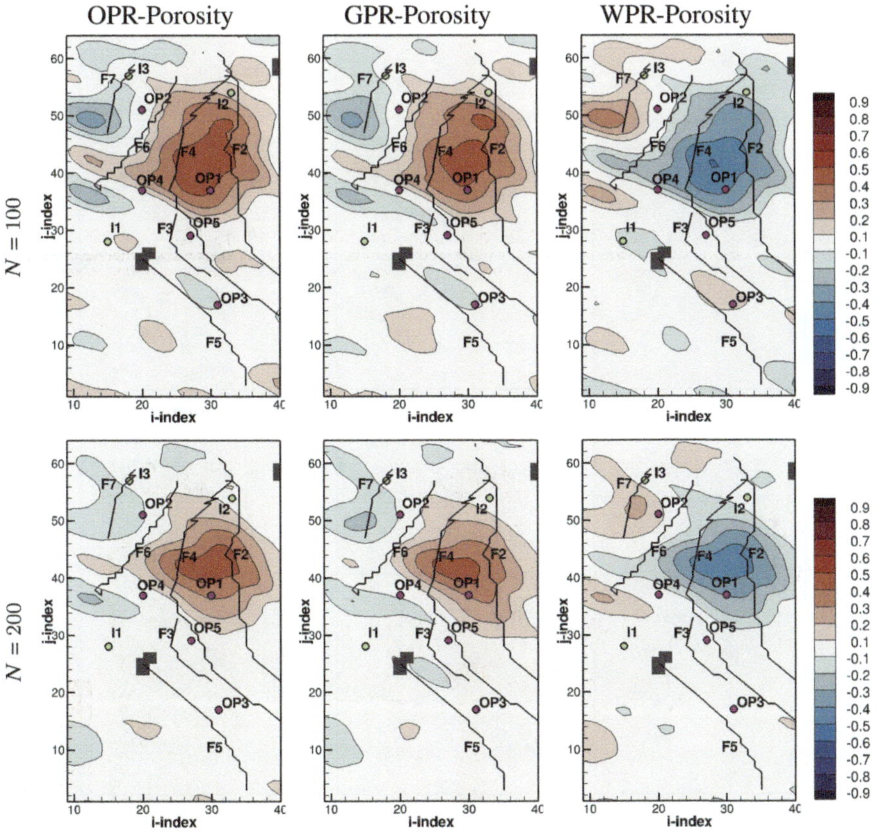

Fig. 13.6 The panels show from left to right the ensemble oil-, gas-, and water-production rate correlations with porosity for rate measurements taken in month 25 of the simulation. The upper panels used an ensemble size of $N = 100$, and the lower panels used $N = 200$

ensemble of $N = 100$, $N = 200$, $N = 400$, and $N = 1000$ realizations and show how they become less contaminated by sampling errors the larger the ensemble size. The magnitude of the spurious correlations fits well with the theoretical values for the various ensemble sizes. In the 100 realizations case, we have spurious correlations exceeding 0.3, and with 1000 realizations, there are no spurious correlations larger than the contour interval of 0.1.

We note that the maximum physical correlation is around 0.45 in this case. If we directly observed the porosity variable, we would have had a maximum correlation equal to one. However, we obtain a lower maximum physical correlation in the reservoir case where the rate data connects to a nonlinear spatial average of petrophysical fields (Li et al. 2003). This fact has a significant impact on our ensemble HM system. In the $N = 100$ case, we have spurious correlations more prominent than 0.3, i.e., of similar order as the maximum physical correlations. Doubling the ensemble size

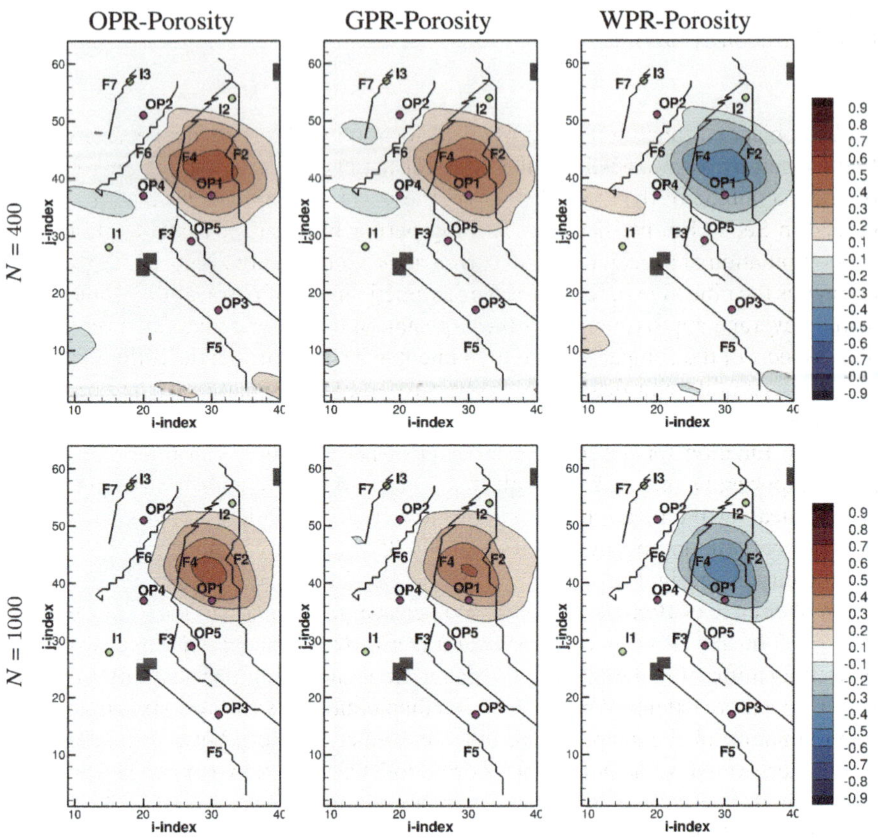

Fig. 13.7 The panels show from left to right the ensemble oil-, gas-, and water-production rate correlations with porosity for rate measurements taken in month 25 of the simulation. The upper panels used an ensemble size of $N = 400$, and the lower panels used $N = 1000$

to $N = 200$ significantly reduces the spurious correlations. They are now mostly less than 0.21, and with $N = 400$ realizations, we lessen the spurious correlations even further to be below 0.15. Thus, with smaller ensemble sizes, we can expect significant updates in the porosity field throughout the model domain due to spurious correlations. These plots clearly illustrate that it is vital to use an ensemble size large enough to distinguish between physical correlations and spurious correlations, and it seems that this issue is more critical in petroleum applications than what one has seen in other geoscience applications, where one assimilates direct observations of the state variables.

An interesting phenomenon is that correlations extend across closed faults. Thus, the flow regime alone does not determine the shape of the correlation functions. Remember that the original depositional environment was smooth and correlated before being faulted, and these correlations will persist even after the faulting.

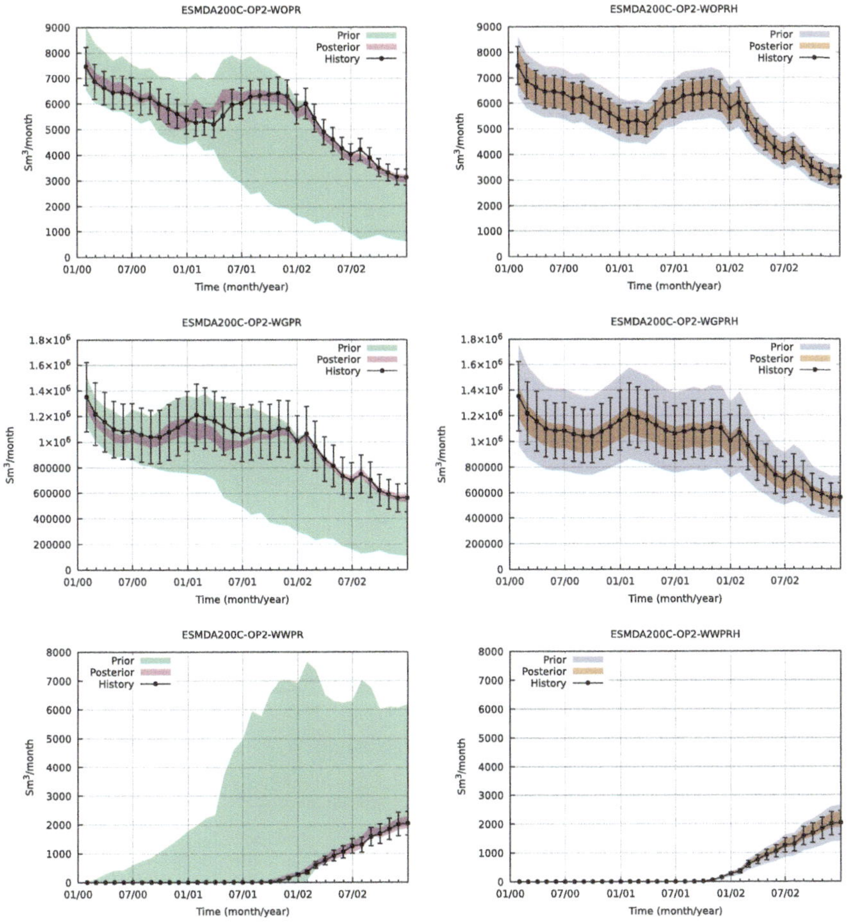

Fig. 13.8 ESMDA200C: $N = 200$, global analysis conditioning on time series of rates using a diagonal $\mathbf{C}_{dd}$ and including control uncertainty

Note also that the correlation function corresponding to well P1 is not centered at the well location but is shifted towards the injector I2 and covers most of the area between the injector and the producer. We expect a significant flow from the injector to the producer that determines the production in P1. Thus, the reservoir properties between P1 and I2 are the main constraints for the oil production in P1. Any change to the reservoir properties on the opposite side of P1 will not affect the production rates, and the data collected at P1 does not provide any information about the reservoir there. The region of significant cross-correlation between the observation and the reservoir property is determined jointly by the correlation length of the prior model covariance matrix and the area of data sensitivity to model parameters (Chen and Oliver 2010b; Emerick and Reynolds 2011). This example shows that the ensemble correlation

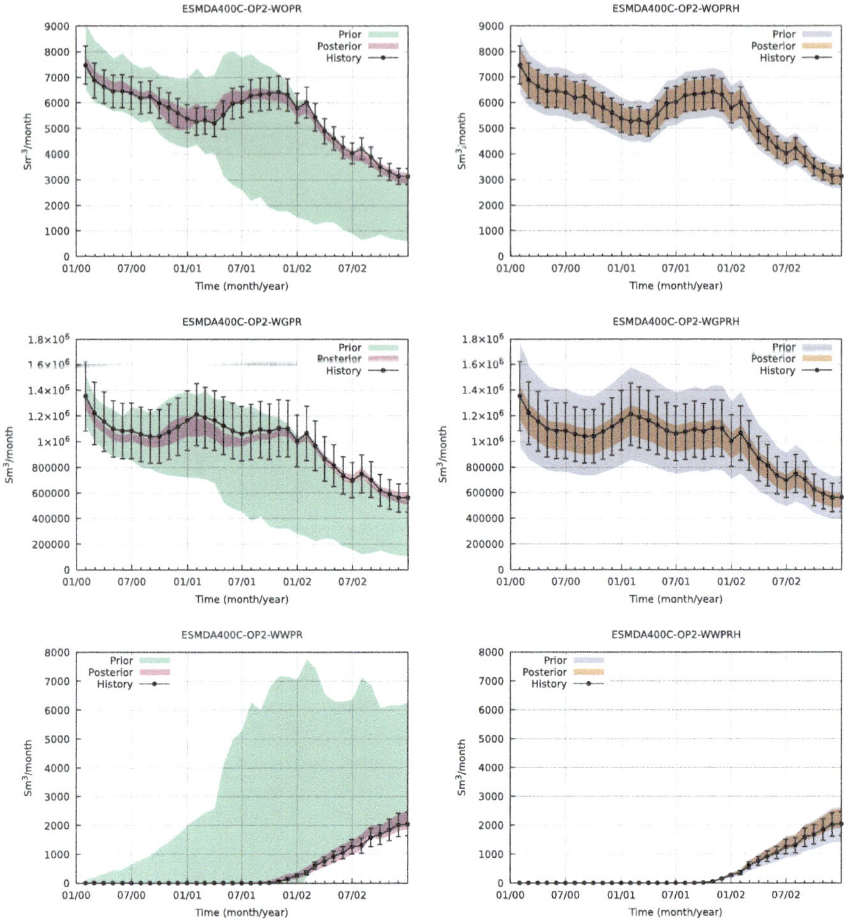

Fig. 13.9 ESMDA400C: $N = 400$, global analysis conditioning on time series of rates using a diagonal $\mathbf{C}_{dd}$ and including control uncertainty

function is not centered on the observation location when the measurements (injection and production rates) are primarily sensitive to the properties between an injector and a producer. For this reason, one sometimes shifts localization regions in distance-based localization from the measurement location (Chen and Oliver 2010a).

We also show the production rates from the cases ESMDA200C, ESMDA400C, and ESMDA1000C in Figs. 13.8–13.10.

Another vital contribution to the violent porosity updates is the negligence of dependencies in the historical rate data time series, which we will discuss next.

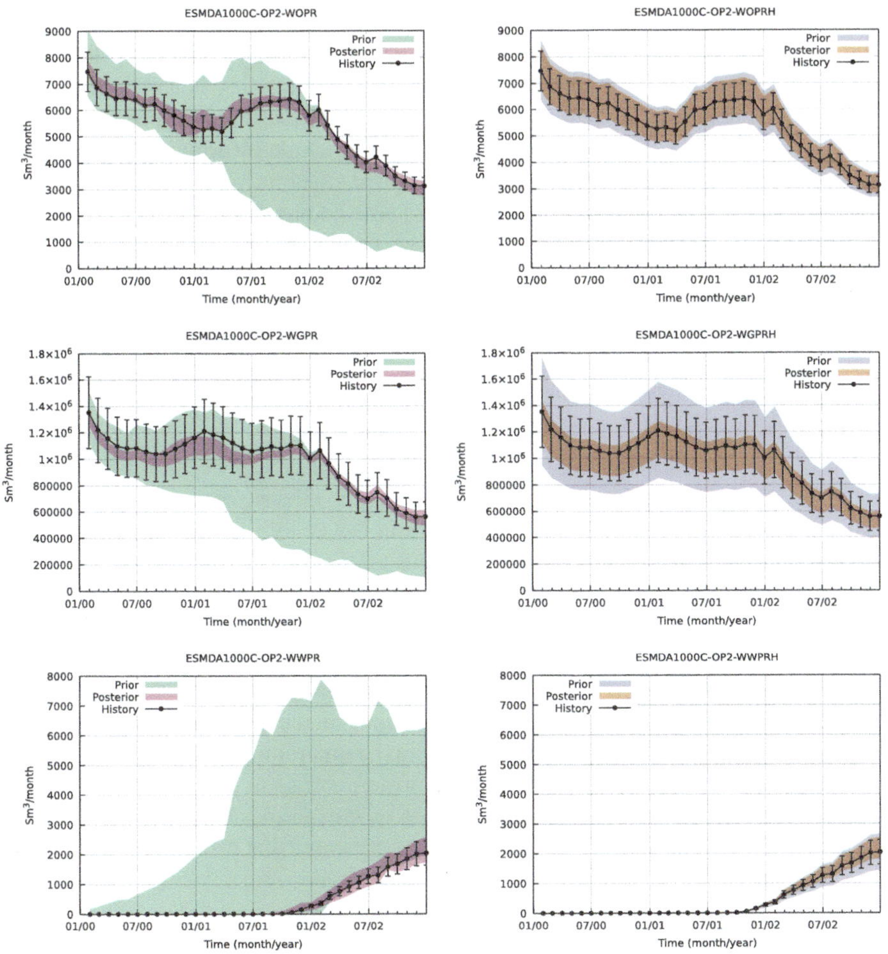

Fig. 13.10 ESMDA1000C: $N = 1000$, global analysis conditioning on time series of rates using a diagonal $\mathbf{C}_{dd}$ and including control uncertainty

13.6 Redundancy of Historical Rate Data

In an example with the Norne reservoir model, Evensen and Eikrem (2018) showed that *it is possible to obtain an equally good match to the historical rate data when conditioning the model only on the total accumulated production at the end of the historical period for each well as when including the whole time series of rate data in the conditioning*. In this case, we will only have three measurements per well, i.e., the total accumulated oil, gas, and water production. The rationale behind this result is that a given pressure difference between an injector and producer will drive a diffusive flow toward the producer, resulting in a particular production rate that depends on the reservoir properties. We can assume that the flow is proportional to the pressure difference. Therefore, a single rate measurement explains the relation between production rate and pressure difference. Adding a time series of such rate data will likely contribute redundant information since the reservoir properties do not vary with time. In some cases, a time series will include information regarding the timing of a water breakthrough or other events that the total accumulated production cannot identify. Evensen and Eikrem (2018) also pointed out that the typical erratic behavior seen in time series of rate data is often a result of how the production engineers control the wells. It is not a result of model nonlinearities or turbulent or chaotic flow in the reservoir, as the reservoir flow is typically slow and "boring."

But why would it be beneficial to subsample the rates, even to the extreme extent of representing the whole time series by only three total accumulated production values per well?

Given that a time series of rate data contains primarily redundant information, we do not gain much by adding another rate measurement to the time series. For example, will we provide more information about the reservoir flow if we switch from monthly to weekly historical rate data in the conditioning? The consequence of the measurement redundancy is that the product $\mathbf{S}\mathbf{S}^{\mathrm{T}}$ in Algorithm 1 will be singular and of low rank. Suppose we assume uncorrelated measurement errors by setting $\mathbf{C}_{dd} = \mathbf{I}$. In that case, it means we are adding the same information repeatedly with each rate measurement, which leads to a collapse of the ensemble variance. This situation resembles a case where we observe the temperature at a location many times using independent thermometers, and by averaging all the measured temperatures, we will obtain a very accurate estimate of the actual temperature. So, we should be aware that we will get a very low variance in the posterior ensemble from conditioning on a time series of redundant rate information. As the examples below show, increasing the ensemble size or using localization will not resolve this problem.

Note also that the problem with redundancy and dependency occurs even in synthetic cases where a reference solution provides a "true" prediction. The rate data provide redundant information, and if we assume uncorrelated errors, this corresponds precisely to the case with many independent temperature measurements. The result is a very low posterior ensemble variance, which is correct given the specified prior information.

On the other hand, in real reservoir problems, the time series of rate data are strongly dependent, i.e., they have correlated errors. This fact becomes apparent when considering how the data are "generated." We use the word generated, even though measurements usually are measured. Typically, the rate data are not measured directly but instead read out of an allocation table, e.g., given input parameters such as valve settings and strength of the gas lift. Hence, any error in the allocation table will contaminate the rate data in a time series the same way, leading to highly correlated errors in time or, equivalently, a bias. The allocation tables are typically only updated every few years, depending on the reservoir. Thus, we must account for the error correlations within each "allocation window" when computing the assimilation update. Suppose we wrongly assume that measurement errors are uncorrelated. In that case, the measurements will have a too-large impact on the update, leading to excessive updates and underestimated posterior ensemble variance, as was shown in Sect. 6.7. We have seen this effect in the previous sections when computing the global updates based on the time series of presumed independent rate data. Note the work by Evensen (2021), who used proper error statistics for the rate data, leading to consistent results.

We obtain the correct result for synthetic cases with a very low posterior variance when conditioning on a time series of independent rate data. On the other hand, a more plausible case setup would be using a single or very few independent rate data. Fortunately, in realistic reservoir cases, we know that the rate data are dependent, and we should include this knowledge in the prior by specifying the correct measurement error correlations. Alternatively, we can subsample the rate data or condition only on the total accumulated production for each fluid.

A significant problem in most operational data assimilation implementations is that they, for simplicity, employ the "exact-inversion" algorithm from Sect. 6.6 that *implicitly assumes measurement errors are uncorrelated.* Thus, a subsampling of the rates becomes vital to avoid ensemble collapse and to obtain a correct result. However, we point to the subspace inversion by Evensen (2003), see Sect. 6.5, which computes the inversion in the ensemble subspace to a cost equivalent to the "exact inversion," and allows for specifying a full $\mathbf{C}_{dd}$. Thus, the "exact-inversion simplification" is unnecessary, and we could employ the subspace inversion algorithm from Sect. 6.5 with a time series of rate data and an appropriately prescribed $\mathbf{C}_{dd}$. Evensen et al. (2024) also found that the subspace inversion from Sect. 6.5 may be more numerically stable than the commonly used exact inversion from Sect. 6.6.

In the following experiments, we condition the model on only 15 observations, i.e., the total accumulated oil, gas, and water production at the end of the historical simulation for each of the five wells. This approach is a "poor man's" remedy for dealing with the error correlations in the time series of rate data. There are only three measurements per well, and we do not need to concern ourselves with the highly correlated errors in the time series of rate data.

In Fig. 13.11, we show the results from a global analysis using 100, 200, 400, and 1000 realizations and conditioning the total accumulated production. It is clear that the reduction in the ensemble's standard deviation is now much smaller when we condition only on the total accumulated production, and the reason is that we are not adding multiple "copies" of the same information again and again from a long time series of rates. The updates of the porosity field become much weaker and probably more realistic when we condition the model on the accumulated rates. When we increase the ensemble size to 1000 realizations, the updates become slightly smaller and more localized around the production-well locations as we have reduced the impact of spurious correlations.

Thus, we conclude that the conditioning on time series of dependent data while ignoring their error correlations, and with spurious correlations resulting in sampling errors, leads to excessive updates and "ensemble variance collapse." A large ensemble will reduce the impact of spurious correlations and improve the results, but it will not fix the problem with neglected measurement error correlations.

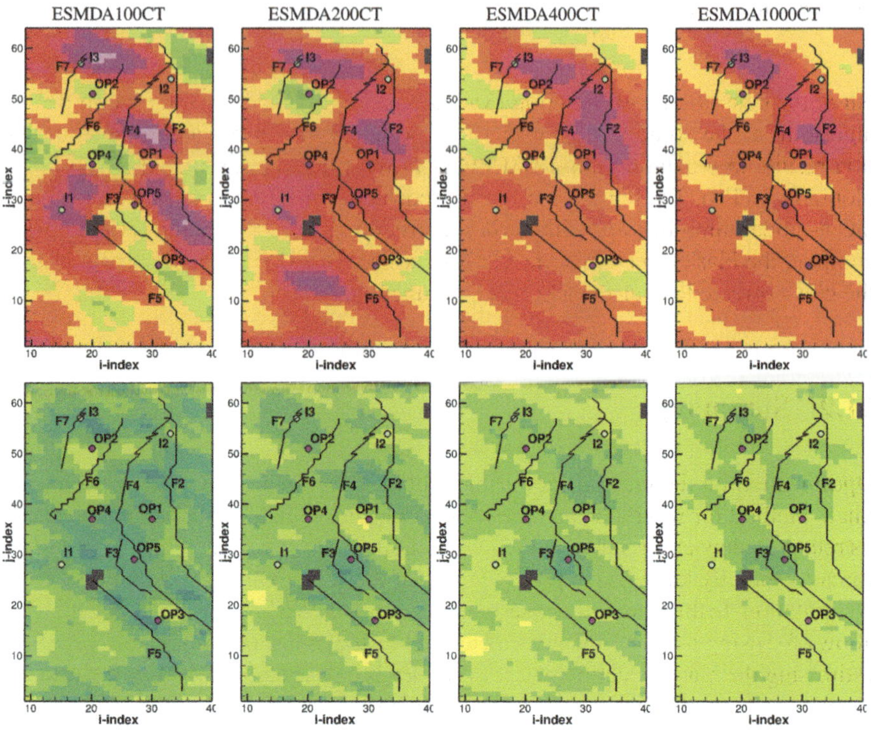

Fig. 13.11 Conditioning on accumulated production: Global updates using different ensemble sizes as in Fig. 13.5, but conditioning on the total accumulated production

In addition to the much weaker updates, we note that from Figs. 13.12–13.15, we now retain a much more significant uncertainty in the predicted rates similar to the magnitude of the observed rate errors. Still, we note that we underestimated the ensemble spread in the case of 100 realizations. Since we will always be limited in the number of model realizations, we must deal with spurious correlations in some other manner. The standard approach uses a localization method, and we will continue the discussion using a correlation-based adaptive localization in the next section (Figs. 13.13 and 13.14).

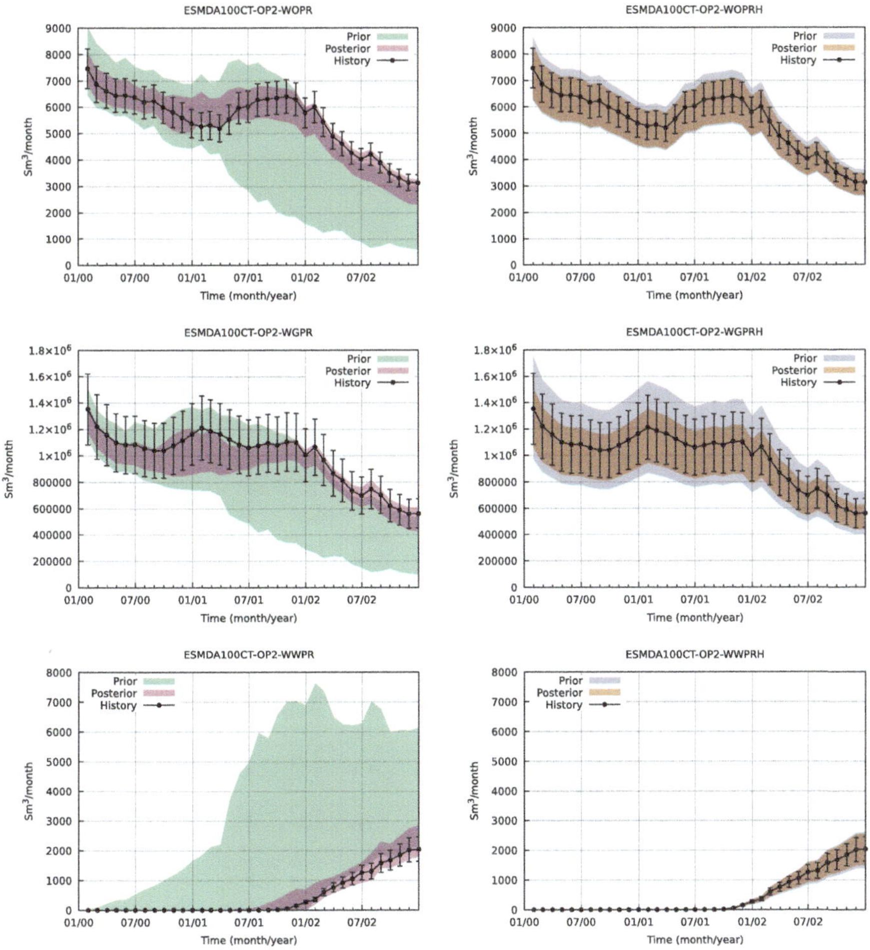

Fig. 13.12 ESMDA100CT: $N = 100$, global analysis conditioning on total accumulated rates and including control uncertainty

13.7 Localization When Conditioning on Time Series of Rate Data

We will now examine the impact of correlation-based localization when conditioning the model on the historical rate data. We have run three cases using $N = 100$, $N = 200$, and $N = 400$ realizations with the corresponding truncation values determined by the theoretical truncation values 0.30, 0.21, and 0.15 from Chap. 7. In Fig. 13.16, we show the resulting porosity fields for the three cases.

With localization, we retain more of the variance in the posterior ensemble, as we remove a large part of the spurious correlations. However, we still see too strong

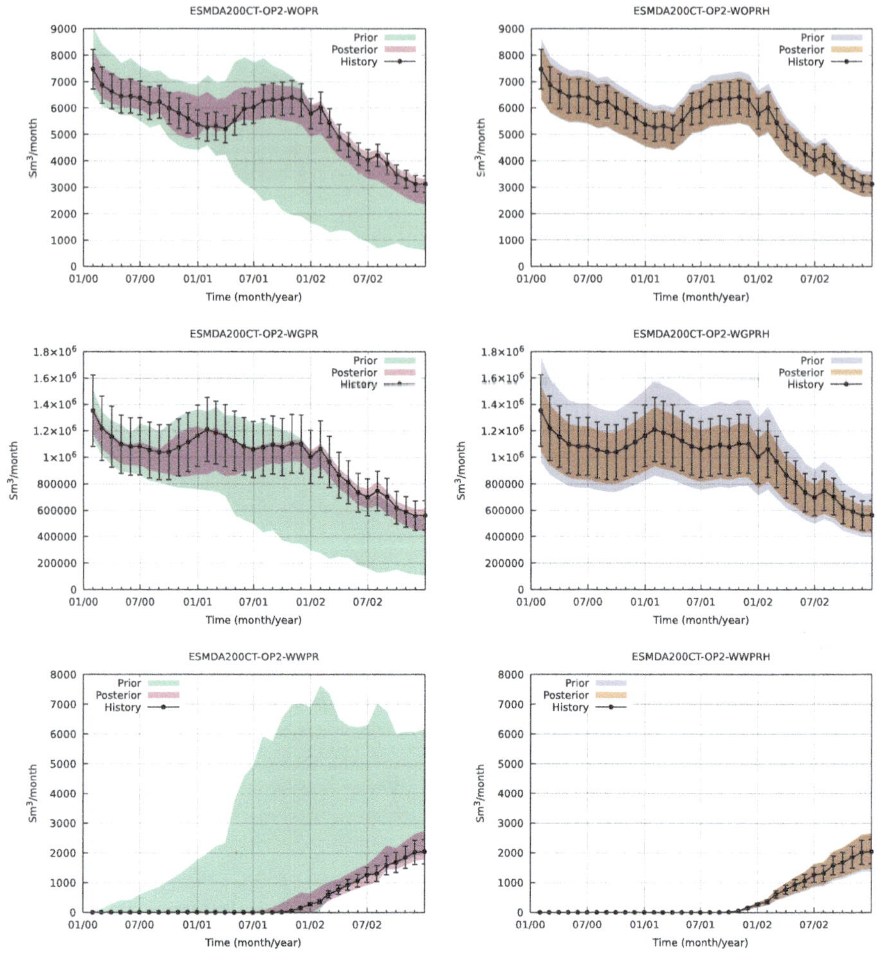

Fig. 13.13 ESMDA200CT: $N = 200$, global analysis conditioning on total accumulated rates and including control uncertainty

updates at the production well locations. The excessive updates are due to the conditioning on the time series of dependent rate data, where we have ignored the dependence by using a diagonal measurement error covariance matrix. We also need 200 or 400 realizations to obtain acceptable results because of the low signal-to-noise ratio with small ensemble sizes (the spurious correlations are of the order of the physical correlations). However, with 200 and 400 realizations, the correlation-based localization does an excellent job of eliminating the impact of spurious correlations.

The match to the historical rates is also excellent for all the localization cases, as seen in Figs. 13.17, 13.18 and 13.19, but we still underestimate the posterior variance. The cause for the underestimation is that even with the localization, we still retain

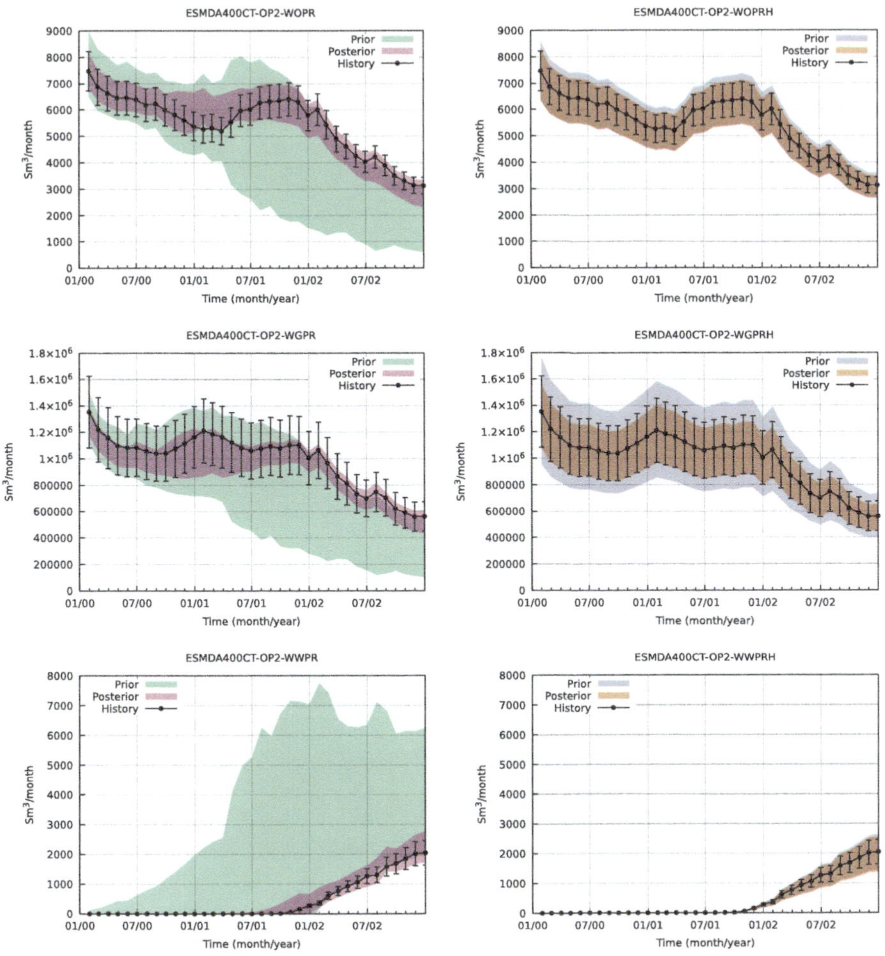

Fig. 13.14 ESMDA400CT: $N = 400$, global analysis conditioning on total accumulated rates and including control uncertainty

the whole time series of rate data in the analysis update. The localization can only filter on the spatial location of the variables we update. Near a well, we will condition the model on the whole time series of rate data. Since we have neglected the error correlations in the time series of rate data, we impose too much information on the update and underestimate the variance.

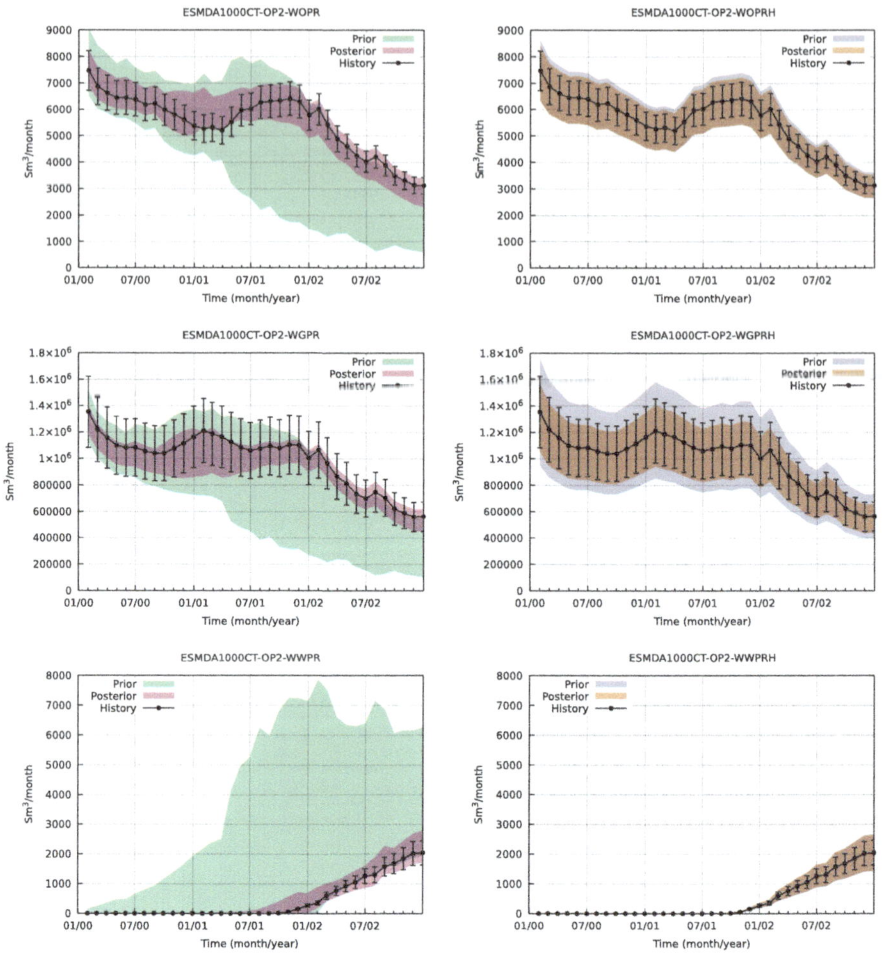

Fig. 13.15 ESMDA1000CT: $N = 1000$, global analysis conditioning on total accumulated rates and including control uncertainty

13.8 Localization When Conditioning on Accumulated Production

We now repeat the experiments from the previous section but combine adaptive localization and conditioning on the total accumulated rates. Figure 13.20 shows the standard deviation and mean for three experiments using different ensemble sizes. Both localization and the conditioning on total accumulated production reduce the strength of the update. From Figs. 13.21, 13.22 and 13.23, the corresponding match to the historical rates is excellent in these cases, and we retain a significant and realistic

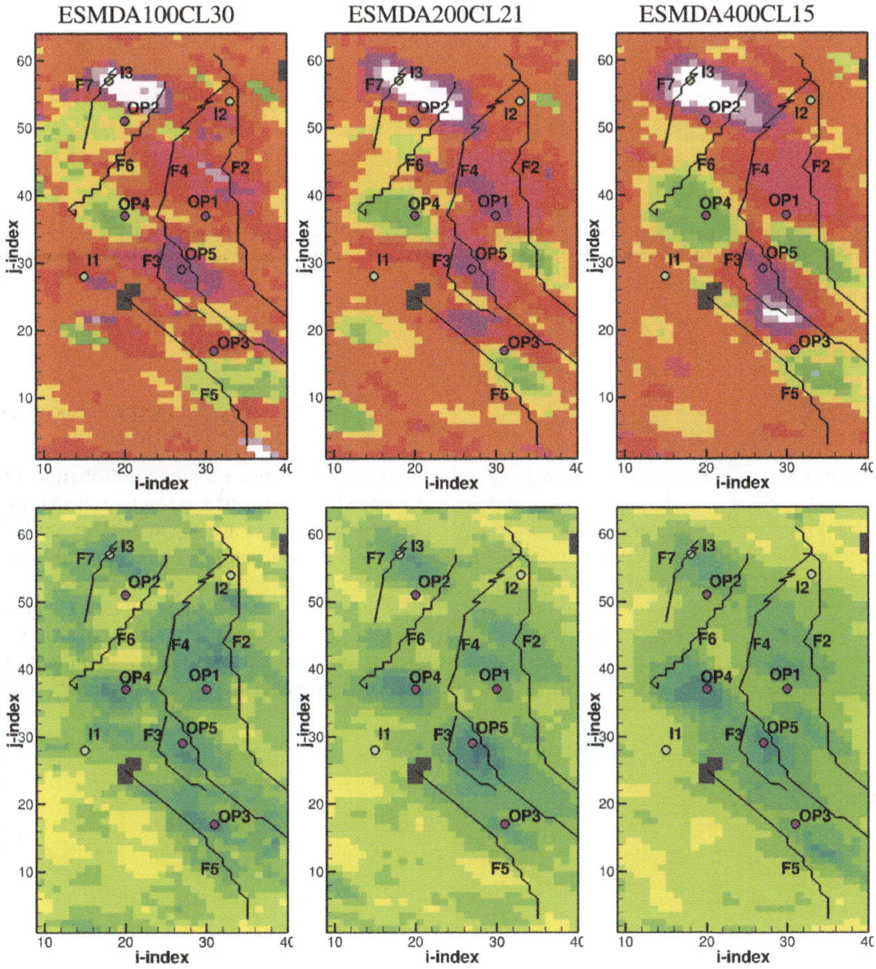

Fig. 13.16 Local update: The plots show the porosity ensemble average (top) and the porosity ensemble standard deviation (bottom) for layer four

uncertainty in the ensemble predictions. Thus, we have computed a minimum update to the prior that provides an acceptable match to the rate data.

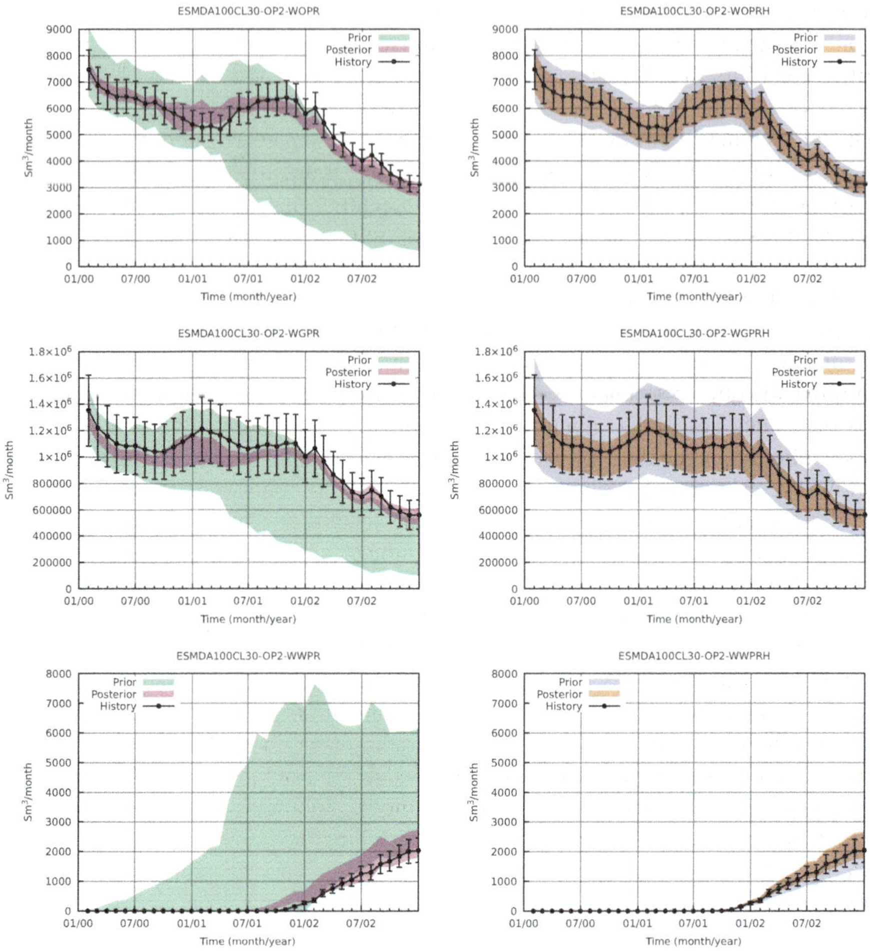

Fig. 13.17 ESMDA100CL30: $N = 100$, local analysis conditioning on full-time series of rate data

13.9 Summary and Recommendations

In this chapter, we have comprehensively illustrated how to formulate and solve the reservoir history-matching problem. We started with the standard case, where we updated only the static model parameters and neglected error correlations in the time series of rate data on which we conditioned the model. This configuration leads to a strong underestimate of the posterior ensemble variance (almost an ensemble collapse).

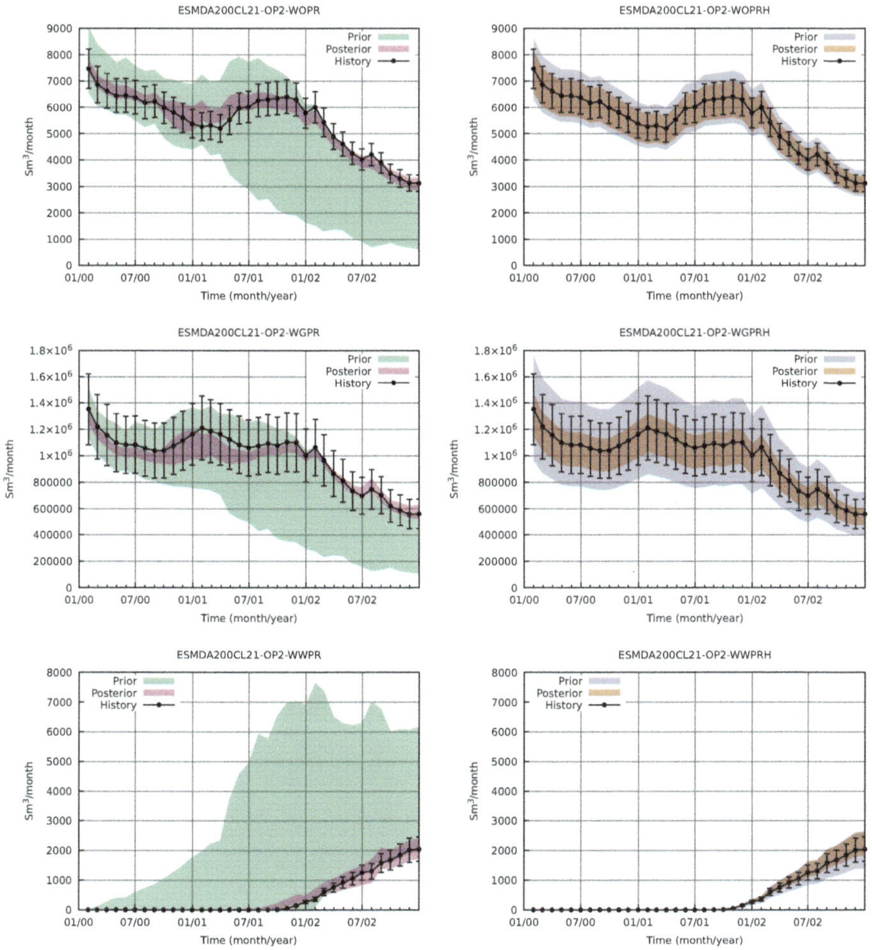

Fig. 13.18 ESMDA200CL21: $N = 200$, local analysis conditioning on full time series of rate data

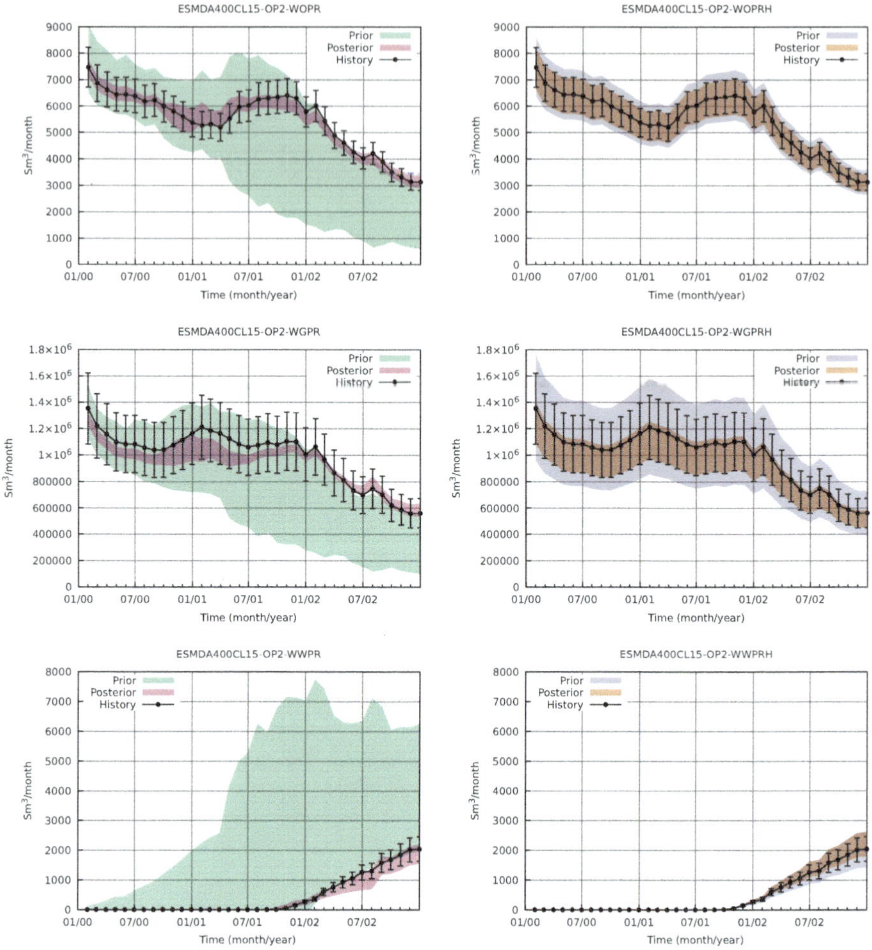

Fig. 13.19 ESMDA400CL15: $N = 400$, local analysis conditioning on full time series of rate data

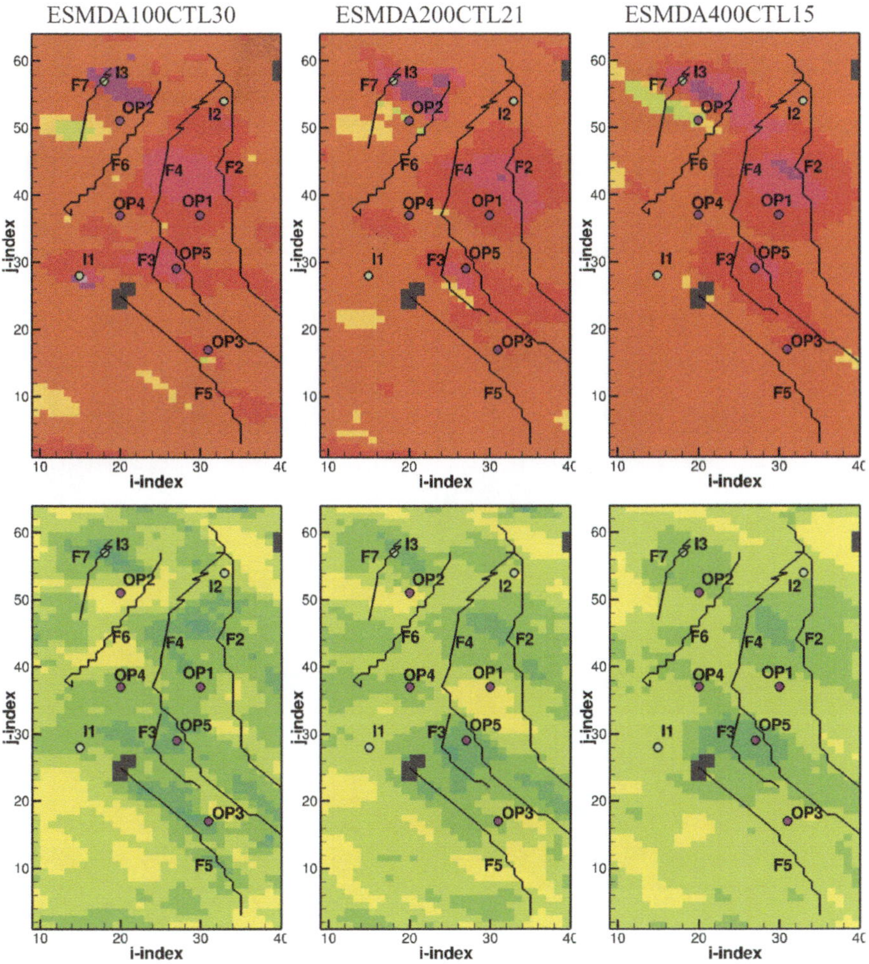

Fig. 13.20 Using localization when conditioning on accumulated production: The plots show the porosity ensemble average (top) and the porosity ensemble standard deviation (bottom) for layer four

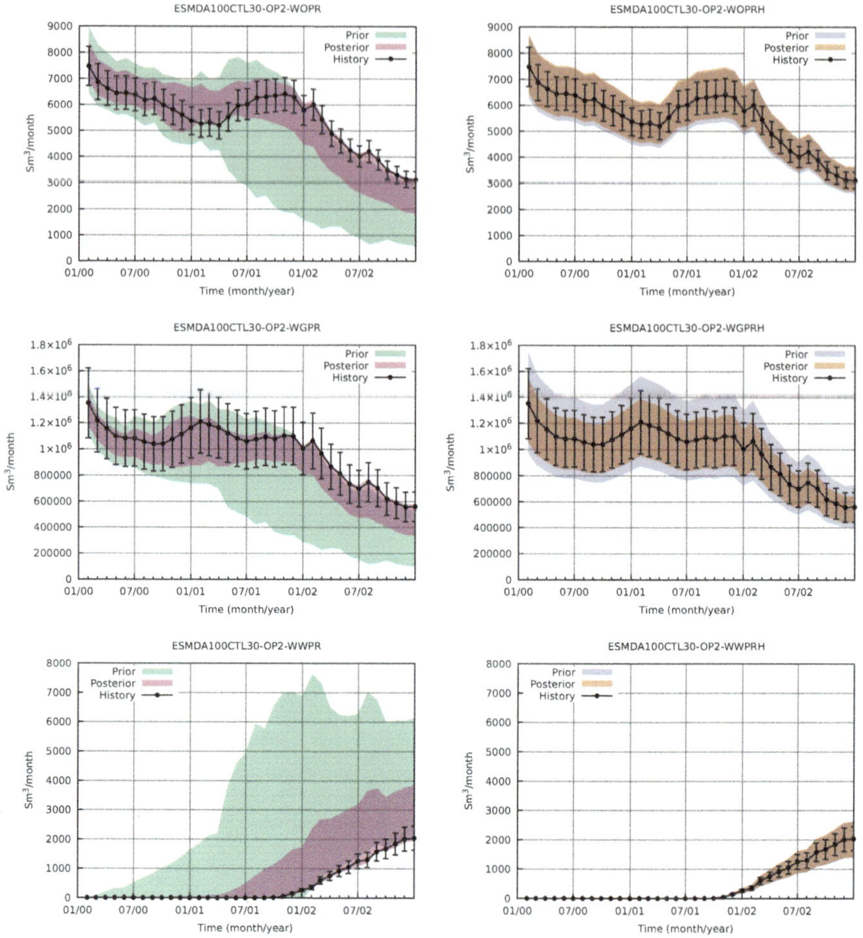

Fig. 13.21 ESMDA100CTL30: $N = 100$, local analysis conditioning on total accumulated rates and including control uncertainty

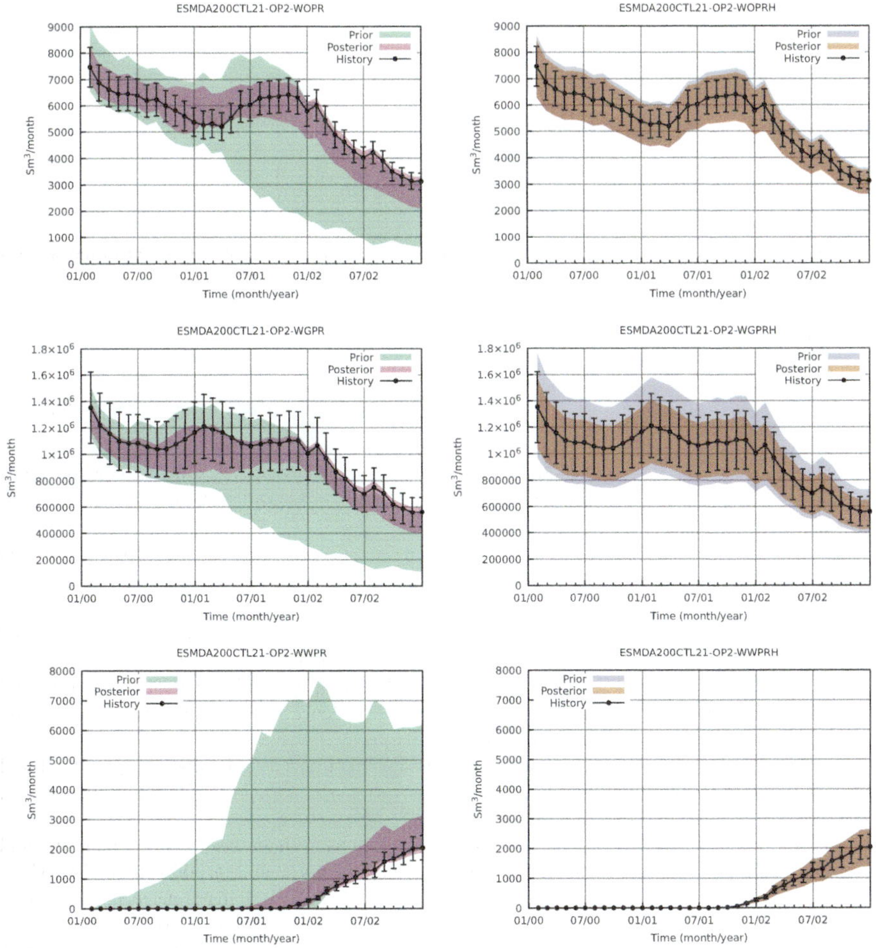

Fig. 13.22 ESMDA200CTL21: $N = 200$, local analysis conditioning on total accumulated rates and including control uncertainty

Next, we demonstrated the importance of including the uncertainty of historical rates when we use them to force the ensemble integration. By including these uncertain controls in the state vector and updating them like we update uncertain model parameters, we ensure a more realistic prior and posterior ensemble prediction for the production-well rates with higher variance.

We have also illustrated that the spurious correlations with small ensemble sizes, e.g., $N = 100$, have almost the same magnitude as the physical correlations between historical rates and petrophysical variables. In the current example, the maximum physical correlation between the porosity field and the production rates is between 0.4 and 0.5. It is lower than one because of the nonlinear relationship between the

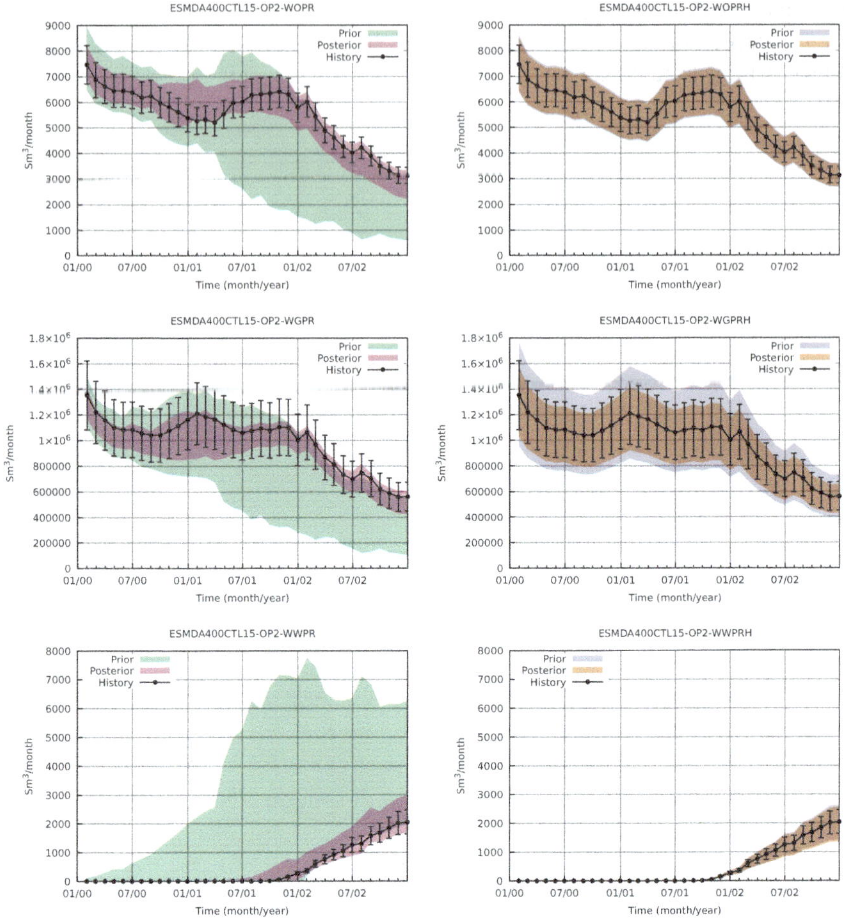

Fig. 13.23 ESMDA400CTL15: $N = 400$, local analysis conditioning on total accumulated rates and including control uncertainty

porosity and the production rates. Thus, it is necessary to use an ensemble size of order $N = 200$ to ensure a significant separation between physical and spurious correlations. These numbers can be case-dependent, and we should verify them from case to case.

We have also shown that correlation-based localization works well if the ensemble size is large enough. In particular, with $N = 200$, we have a good separation between physical and spurious correlations, and a correlation truncation at $3/\sqrt{N} = 0.21$ works well. The localization significantly reduces the updates far from the production-well locations. However, suppose we include the whole time series of rate data from a well when updating variables. In that case, it becomes essential to correctly specify the measurement error statistics to avoid underestimating the posterior ensemble variance.

As an "alternative" to specifying a total measurement error covariance matrix, we ran examples in which we conditioned only on each well's accumulated total oil, gas, and water production. This approach resembles an effective subsampling of the rate data. Since the information provided by a time series of rate data contains much redundant information, we still obtain an excellent match to the historical rates. Still, when using global analysis, we have a significant impact of spurious correlations, which deteriorate the results far from the production-well locations.

The combination of correlation-based localization and conditioning on the total accumulated production resulted in the "best" results. By best, we define the minimal update to the prior ensemble of parameters that results in an acceptable match to the historical rate data and with a realistic uncertainty.

Evensen (2021) obtained similar results when he conditioned the model on the entire time sequence of historical rates but with full specification of the error correlations in the measurement error covariance matrix.

A final comment is that all the cases discussed above match the historical rates, where the results differ mainly in the posterior variance. Thus, while it is common to validate the history-matching experiments concerning the match to the historical rates, it is also crucial to validate the realism of the updates to the prior parameters. An example is the violent updates we obtain for the porosity field when computing the global update with $N = 100$ realizations.

Chapter 14
History Matching the Troll Reservoir

Abstract This chapter presents a vast and complex reservoir history-matching case for the Troll field on the Norwegian Continental Shelf. It follows the same history-matching procedure as in the previous chapter. The Troll examples serve as a further confirmation of the learnings and recommendations coming out of the REEK experiments, and the examples illustrate the properties of the history-matching problem when conditioning on production rates. In addition, it uses the adaptive pluri-Gaussian method for parameterizing facies uncertainties and updating these in the history matching. The examples start with a standard history-matching procedure, computing global updates. After that, examples illustrate how the results improve when one properly accounts for uncertainties in the model controls and the facies parameterization. Finally, an adaptive localization scheme reduces the impact of spurious correlations and significantly improves the history-matching results.

14.1 Introduction

Reservoir[1] characterization[2] is an approach for modeling and representing a reservoir to understand its geological structure, petrophysical properties, and fluid flow regimes. It combines static and dynamic subsurface data as input for subsurface modeling tools and numerical flow simulators. This process integrates numerical simulations with production data information to model subsurface fluid saturations, flows, and pressure distributions. The resulting subsurface model forms the basis for reservoir management and aids the decision-making process when developing production strategies, including field development and well-planning. Additionally, the asset team uses the reservoir model to estimate reserves and generate production

[1] The simulation and history-matching work presented in this chapter was performed in Equinor by Kjetil Bjørke and Torbjørn Ek from the Troll asset team in collaboration with the developers' support team for FMU and ERT, represented by Feda Curic. The book's authors wrote the text, which presents the experimental setup, the methodological approach, and the analysis of the results.

[2] We must point out that the conditioning on total accumulated production is not yet recommended practice in Equinor, and the Troll experiments were conducted purely for research purposes.

G. Evensen et al., *Ensemble History Matching*,
https://doi.org/10.1007/978-3-031-99155-4_14

forecasts. Some more applications involve developing improved oil recovery (IOR) plans, greenhouse gas emissions control, and carbon capture and storage (CCS) planning.

For the Troll Field, reservoir characterization has been integral from the beginning. Simulation studies have played a key role in evaluating and developing the *plan for development and operations* (PDO) and making decisions regarding well placement, drainage strategy, and more. The complexity of the geological and associated reservoir simulation models has grown in tandem with advancements in software, hardware, data acquisition, drilling technology, and the accumulation of production data. Ensuring consistency between the reservoir simulation model and the static subsurface model has been challenging due to the intricate geology of Troll and the numerically complex, time-consuming nature of the reservoir simulations.

The following section gives a brief overview of the Troll field with its history and geological concept. After that, Sect. 14.3 briefly discusses the fast model update workflow that allows for efficient history matching and updating of geological input parameters. Section 14.4 presents the history-matching experiments and discusses the results. In addition, it discusses the current state of full-field ensemble simulation models, the alignment between the geological and the reservoir simulation models, and, ultimately, the alignment between the reservoir simulation model and the data observed in the field. The latter represents the conditioning of the geological models on dynamic data in the assisted history-matching process.

14.2 The Troll Field

The Troll field is a vast gas and oil field, covering more than 700 square km in the North Sea, off the western coast of Norway, that started production in 1995. The reservoir comprises Upper Jurassic shallow marine sandstone with excellent reservoir quality (Dreyer et al. 2005). Structurally, three fault blocks dominate the reservoir, which contains a large gas cap, an extensive active aquifer, and a thin oil zone initially 4–27 meters thick. The faults are non-sealing, so local pressure and saturation changes have global effects.

The reservoir's oil is mainly produced through sub-sea horizontal and multilateral wells. The initial plan included 20 horizontal wells in 1995 and 96 horizontal wells in 1996. The complexity and density of wells have significantly increased since then, thanks to advances in drilling and completions technology. Since 2022, the field has been produced by multilateral horizontal wells, mostly with 3–4 branches and some with up to 5–7 branches. A network of wells, manifolds, flowlines, and risers enable the transport of produced oil, water, and gas to the Troll B and Troll C platforms. A cluster of deviated wells below the Troll A platform facilitates gas production.

From the early PDO preparations, reservoir simulations have been essential for understanding the fluid flow in the Troll reservoir. The reservoir model construction now incorporates geological understanding and data, geophysical 3D and 4D-seismic data sets, gravimetric data, and production data. Developing and maintaining the Troll oil-reservoir simulation models is a multi-disciplinary task involving geophysics, geology, reservoir engineering, and production technology. The fast model update (FMU) workflow discussed in the next section facilitates straightforward data integration and model updating. FMU is essential for keeping the model updated with the latest information obtained from production data, 4D-seismic surveys, and the gathering of gravimetric data.

The asset team established three plans for development and operations (PDOs) for the Troll field. Phase 1 focused on gas production from Troll East (TE), Phase 2 on oil production from Troll West, and Phase 3 on gas production from Troll West. Gas production from Troll East began in October 1996. Oil production from Troll West Oil Province (TWOP) commenced in September 1995, followed by gas production from Troll West Gas Province (TWGP) in November 1999 (Madsen and Abtahi 2005). Gas production from Troll West started in 2021. This chapter will focus on the reservoir simulation flow modeling for Troll oil production. The simulation model for oil production must account for pressure changes and fluid flow throughout the Troll region, making the model size a continuous challenge.

Figure 14.1 displays a map highlighting the three main fault blocks in the Troll field consisting of Troll West oil province (TWOP), the Troll West gas province (TWGP), and Troll East (TE). The red dashed line indicates the vertical cross section shown in Fig. 14.2 that passes through the three main fault blocks.

The large extent and thickness of the Troll reservoir and the high number of wells represent a challenge for the geological reservoir modeling. The Troll reservoir contains two very different sand types (Denney 2008): clean coarse sandstone (C sand) with permeability in the range of 1–10 darcys and the finer grain size (M sand) with permeabilities of 1–100 millidarcys. Since the start, the geological modeling approach has been close to a "map-based" (i.e., 2.5D) method with many correlated zones for each C and M sand unit, with interpolated reservoir properties. However, increasing computer capacity makes it now possible to build more detailed geological grids, and the latest geological model comprises around 160 million cells.

Since 2016, the Troll asset team has taken a facies-based approach and almost halved the number of correlated zones in the model. Each zone can have both a C, M, and calcite facies. This facies-based approach leads to greater flexibility in the history-matching loop. It allows the usage of, e.g., adaptive pluri-Gaussian facies methods (APS) by Sebacher et al. (2017), which fits well with the assisted history-matching methods, e.g., the ones used in the Fast Model Update (FMU) workflow, (Hanea et al. 2015).

14.3 Fast Model Update

Fast Model Update (FMU) is an *integrated and automated workflow for reservoir modeling and characterization,* and it is the recommended practice for the modeling and simulation work in Equinor. Considering all relevant uncertainties, FMU connects all steps and disciplines, from seismic depth conversion to prediction and

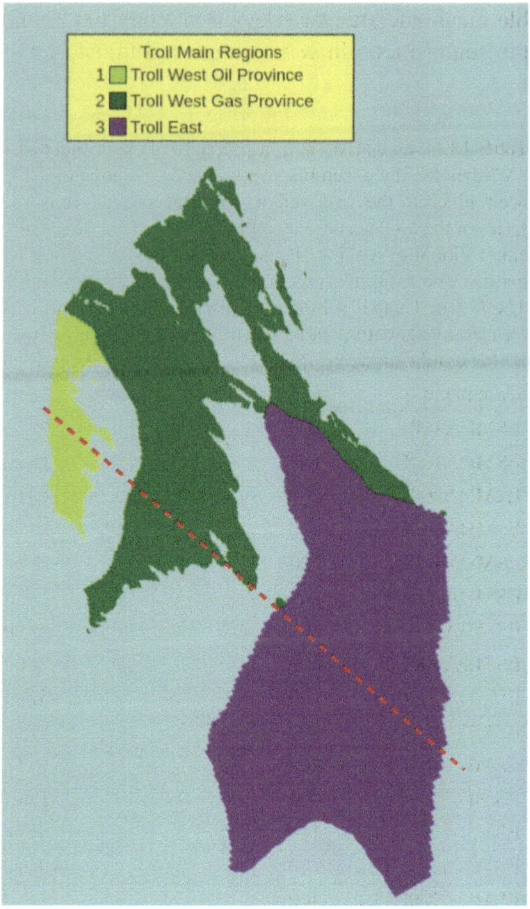

Fig. 14.1 The panel is a map of the three main fault blocks in the Troll field. The red dashed line indicates the cross section in Fig. 14.2. This plot was generated using the RMS software from Emerson and AspenTech

reservoir management. FMU delivers an ensemble of geologically consistent and history-matched model realizations that characterize the reservoir uncertainty. The ensemble history-matching methods represented by Algorithm 1 are the engine of the assisted history-matching process. Hence, the uncertain parameters should be continuous random Gaussian variables. FMU allows updating of the structural model, fault parameters, and petrophysical properties such as 3D fields of permeability and porosity.

Building a reservoir model is a complex task. It involves several disciplines, from seismic processing, interpretation, and time-to-depth conversion to the flow simulation of a coarse simulation model. Geophysicists, geologists, geological modelers, rock physicists, and reservoir engineers contribute to different aspects of the model building. Traditionally, the model-building workflow has been a one-way process, where each step is handled by staff from the corresponding discipline before they hand it over to the following discipline. Some multi-disciplinary collaboration has

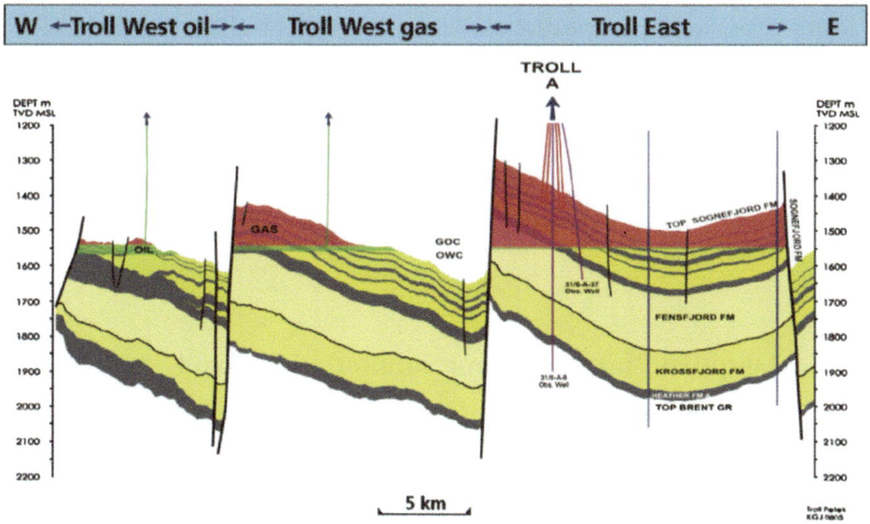

Fig. 14.2 This cross section of the Troll reservoir (reproduced from Fig. 1 in Eiken et al. 2008) passes through the Troll West oil province to the left, the Troll West gas province in the middle, and Troll East to the right (see red dashed line in Fig. 14.1. The depth axes annotate meters below sea level. TVD MSL is the actual vertical depth below mean sea level. The red color indicates gas, and the green denotes oil. The cross section is along the red line shown in the upper panel. Note the location of the Troll A platform for gas production in Troll East

existed, but after the reservoir engineer converges to a history-matched model, transferring the new knowledge to the other disciplines has traditionally been challenging.

With its clever design, the Fast-model-update workflow has transformed the model-building workflow from a stepwise process into an integrated modeling chain. This design ensures the modeling process's consistency, repeatability, and automaticity, showcasing our field's ingenuity and efficiency.

A primary criterion of FMU is the possibility of running the modeling software in batch mode, which allows us to set up many realizations to run in parallel. Such a multi-realization framework is well suited to represent the uncertainty in our input parameters to the model workflow. The uncertainty is represented and propagated through the model-building workflow and flow simulation.

In FMU, changing and adding input parameters is straightforward before running the automated modeling workflow. Another significant upside is that FMU establishes a direct (but nonlinear) link between input parameters and simulated dynamic responses, e.g., oil rates and 4D seismic. Using ensemble methods (EnKF, ES, EnRML), it is possible to analyze the difference between observations and simulated output data and the correlations in the input-output from the model-building workflow. The ensemble methods may use this analysis to decrease the uncertainty span in your prior input parameters significantly.

The facies modeling is an essential step in the model-building workflow. It links the structural framework with the petrophysical properties and distributes the porosity and permeability in the model grid, controlling both the volumes and flow patterns. Several methods exist to simulate the facies realizations, including in an FMU setting. Process-based, object-based, and multi-point statistics can all work. However, until now, there has yet to be a consistent way to use the simulated output and the observed data to condition the facies realizations while honoring the static well-log and seismic data.

14.4 Adaptive Pluri-Gaussian Simulation on Troll

Facies modeling is an essential step in the modeling workflow of the Troll reservoir. The facies are categorical variables, e.g., type C and M sands, and there is a need for a parameterization that projects the discrete variables into continuous ones. Inverted seismic data provides probability cubes for facies distribution in each grid cell of the reservoir model and is the most informative data about the reservoir's facies distribution. Hence, the connection with the probability parameterization of the facies used in adaptive pluri-Gaussian simulations (APS).

Knowledge of the depositional geological environment is key to a successful history-matching process. APS is the method that links the input seismic data cubes, the geological concept, and the truncated pluri-Gaussian method to obtain an ensemble representation of the facies in the reservoir that obeys the well-log data, follows the geology and can be consistently updated. It uses three essential ingredients: the probability cubes, the truncation map, and the Gaussian random fields (GRFs). Recognizing the complexity and required zonation in the Troll model, the APS implementation includes a geological proxy that gives probabilities for the vertical facies distribution at each map location. This proxy is input to a simple APS algorithm that creates "variability" and imposes well-conditioning.

The upper plot in Fig. 14.3 shows the 3D probability proxy model for the clean C sand. Similar proxies represent the M sands and calcites. The lower plot in Fig. 14.3 presents the resulting probability cube, conditioned to the well logs .

The GRF, as shown in Fig. 14.4, represents the geological concept with its trends and statistical properties. The truncation map in Fig. 14.5 describes the location and the interaction between the facies in the reservoir. Following up on the APS workflow presented in the section above, one can simulate a facies type for each model grid cell and in each realization. The lower plot in Fig. 14.4 shows the resulting APS facies model.

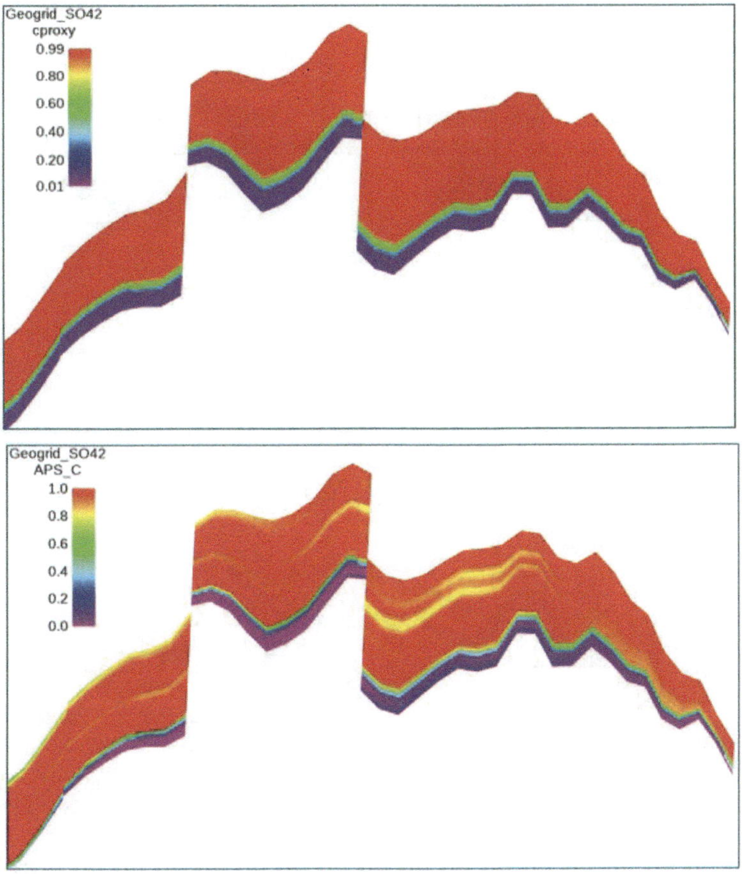

Fig. 14.3 The upper plot shows the 3D probability proxy model for clean C sand. Similar proxies represent M sands and calcites. The lower plot shows the 3D probability proxy model conditioned to wells using petrophysical modeling. These plots were generated using the RMS software from Emerson and AspenTech

14.5 Facies History Matching on Troll

Applying assisted history matching to real-field cases, as opposed to synthetic cases, presents significant challenges due to the complexity of the subsurface model and the considerable uncertainty in the initial reservoir characterization. The primary source of this uncertainty lies in the geological model. Capturing this uncertainty requires simulating a comprehensive set of geological models conditioned on all available subsurface data. Ideally, this set of geological models would encompass all scenarios that fit the subsurface data, incorporating geological concepts such as depositional environment uncertainty and the lateral and vertical extent of deposits,

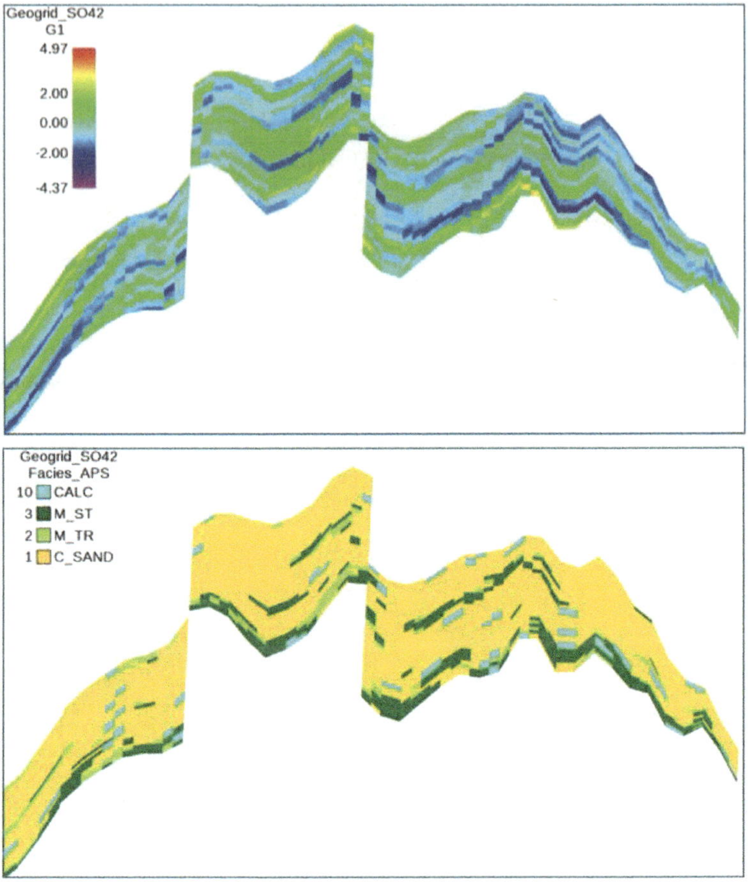

Fig. 14.4 The upper plot shows a Gaussian random field for sands that APS combines with conditioned proxy models to compute the facies distribution. The lower plot is the resulting APS facies model showing the distribution of all facies. These plots were generated using the RMS software from Emerson and AspenTech

structural frameworks with fault networks, facies types, and varying rock and fluid properties.

In Chap. 13, we presented an example of the history-matching setup for a synthetic case, where we tested several aspects of the history-matching formulation. This study provided invaluable findings and conclusions related to the use of a localization method, the type of observations we condition the model on, the specification of model and observation errors, and the choice of ensemble size. These findings are of immense value when working with a large field case where one often has limited computational resources and time. Thus, from the insights gained in Chap. 13, we selected the most successful approach for the Troll history-matching exercise.

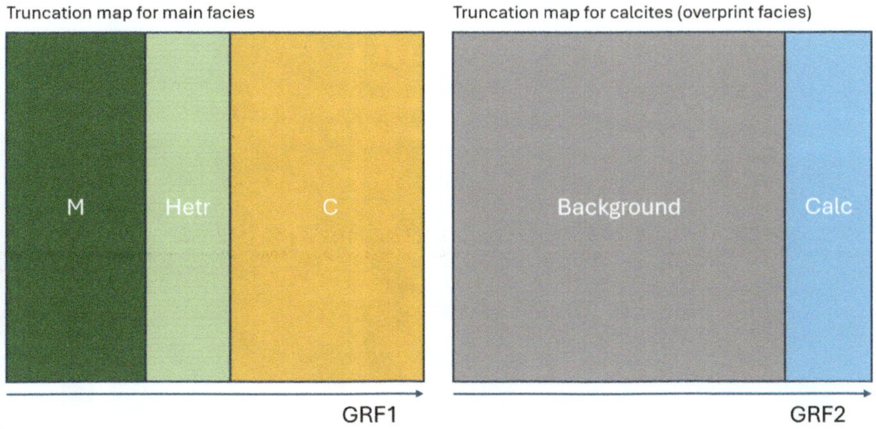

Fig. 14.5 The panels display the truncation map used for the Troll model. It distinguishes between the clean C sand, two types of M sand, and the calcites

This study concentrates on the uncertainties inherent in the facies modeling for the Troll field. The focus is on generating an ensemble of facies realizations conditioned on the production data. We address the challenges of creating multiple geological facies realizations that satisfy all available subsurface data and then the conditioning of these realizations to the dynamical observations, resulting in consistent and realistically updated geological models.

The APS initially produced 200 facies realizations. We recommended using ESMDA for the conditioning process. However, due to the computational cost, where each realization requires approximately 40 h of CPU time, the conditioning used only one iteration of the ESMDA, effectively running the Ensemble Smoother (ES). Our synthetic case study illustrated that applying the adaptive correlation-based localization method with the ES while conditioning the model on the total accumulated production for each well is the most effective method to prevent ensemble collapse and retain significant uncertainty.

In Fig. 14.6, the upper plots display the prior ensemble mean to the left, the posterior ensemble mean from the global update in the center, and the posterior ensemble mean from the local update to the right. The lower plots show the corresponding ensemble standard deviations. It is evident that without localization, the variance of the updated ensemble collapses to zero, indicating that all realizations are nearly identical. In contrast, applying correlation-based localization preserves a reasonable uncertainty in the update.

Thus, as we saw in the REEK case from the previous chapter, when we do not use localization, significant updates occur throughout the model grid, including areas distant from the wells, suggesting the influence of spurious correlations. These intense updates are not consistent with the geological understanding of the reservoir. However, when using adaptive correlation-based localization, the right plots demonstrate how the updates are confined to locations around the wells. This approach maintains

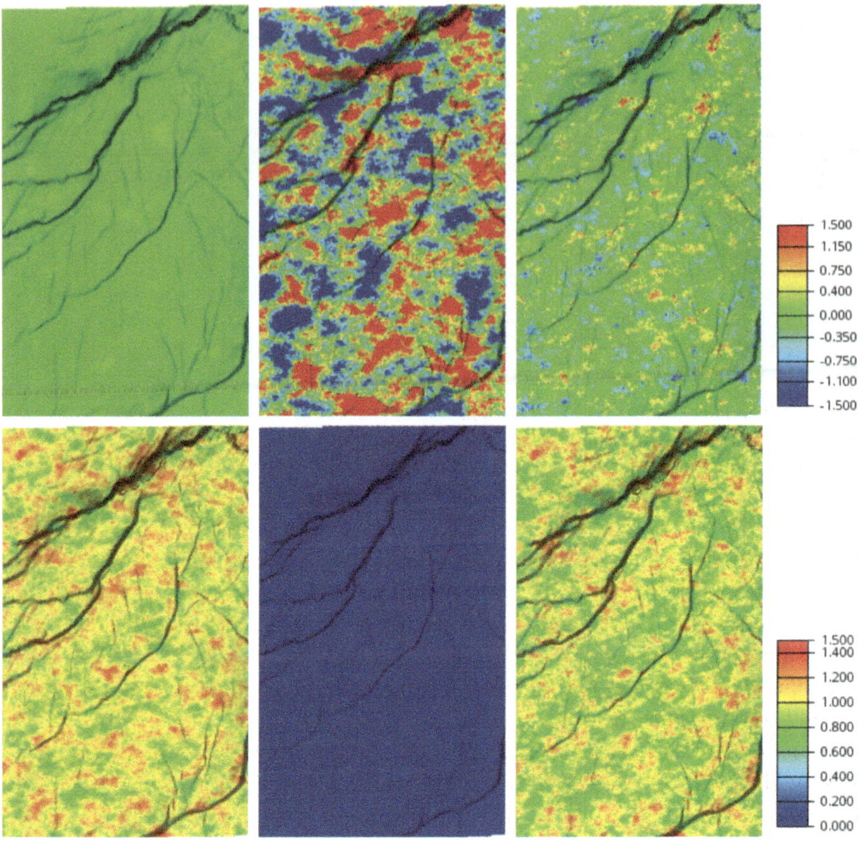

Fig. 14.6 The upper plots show the ensemble mean for the facies: the prior (left), the posterior global update (center), and the local analysis update (right). The lower plots show the corresponding ensemble standard deviations. These plots were generated using the RMS software from Emerson and AspenTech

alignment with the original geological concept, ensuring that updates are consistent with the data without altering the facies distribution in areas not sensitive to the data.

14.6 Conditioning on Production Data

After confirming that the updates align with the initial geological concept and its variability, the next step in assessing the quality of the history-matching results is to examine whether they match the observed production data. We will present the history-matching results focusing on production data matching within the three main sectors of the reservoir, i.e., Troll A, Troll B, and Troll West. Another key aspect is

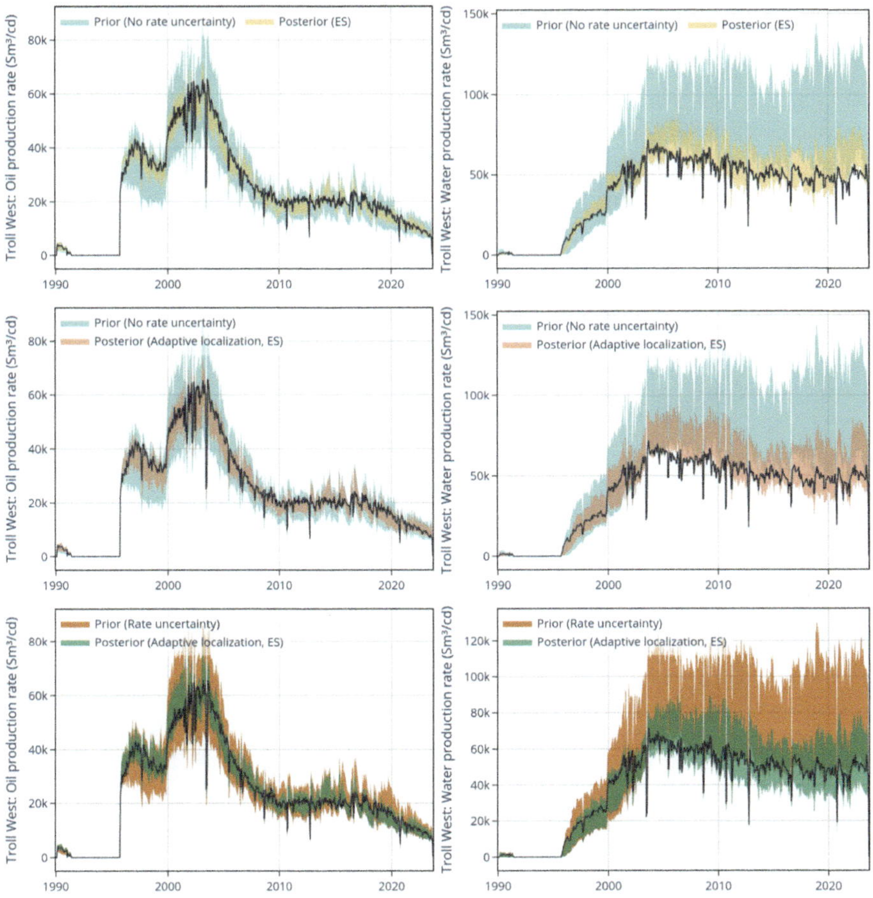

Fig. 14.7 The left and right plots show, respectively, the oil and water production rates for Troll W for the global analysis (top), using adaptive localization but certain rates (second row), and using adaptive localization with uncertain rates (bottom)

comparing simulations, where the focus is on the outcomes of simulations performed with and without incorporating rate-control uncertainties into our model.

In Fig. 14.7, we show the results of the data match before and after conditioning for the oil and water production rates. The first plot shows the results without the rate uncertainty and no localization in the assimilation. The middle plot shows the results that do not account for rate uncertainty but use adaptive localization, and the last one shows the results of the combined rate uncertainty and adaptive localization. The spread of the ensemble in the first plot is minimal, hinting at an ensemble collapse. Also, the updated forecasts are not able to cover the measurements. When using adaptive localization, the ensemble variance becomes more realistic, and the match

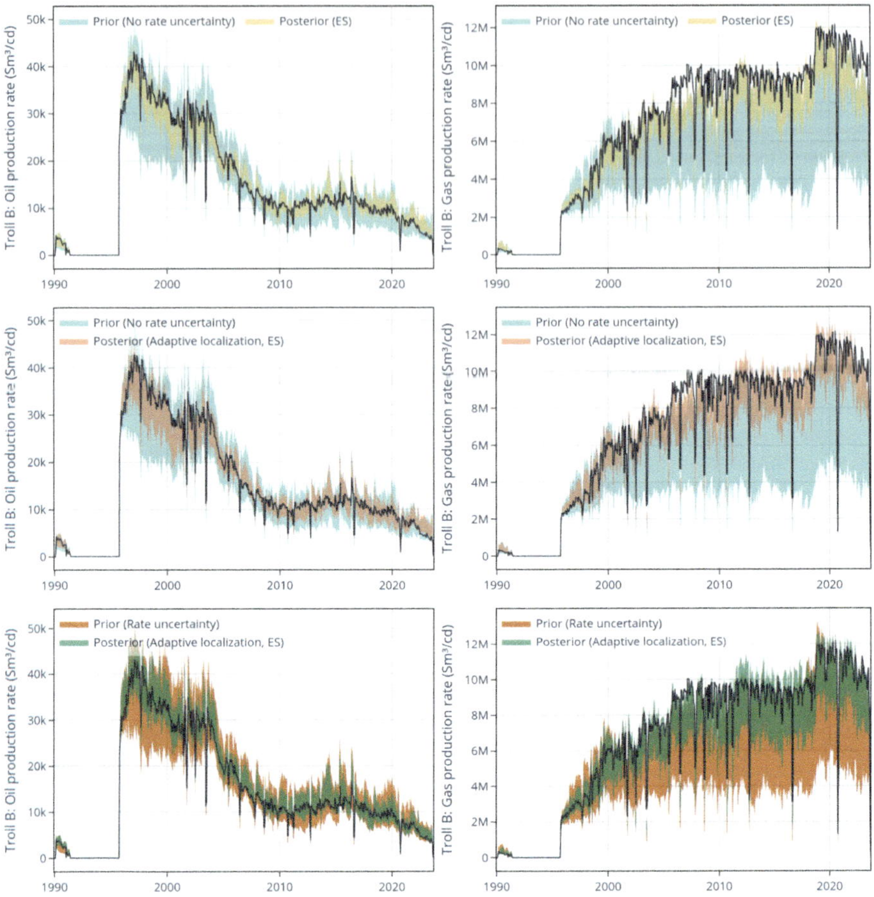

Fig. 14.8 The left and right plots show, respectively, the oil and gas production rates for Troll B for the global analysis (top), using adaptive localization but certain rates (second row), and using adaptive localization with uncertain rates (bottom)

improves. We obtain the best results when including rate-control uncertainties using the adaptive localization in the ES.

We observe the same behavior for the Troll B and Troll C platforms. Figures 14.8 and 14.9 present the results for the gas and oil production rates for Troll B and Troll C. An essential quality check involves assessing the misfits between the predicted and measured observations calculated for the prior and posterior ensembles og geological realizations.

In Figs. 14.10, 14.11 and 14.12, we show the mismatch between the forecasted production rates for oil, water, and gas and the observed data for the prior and posterior ensembles. We have computed the ensemble predictions with and without rate-control uncertainties, and with and without adaptive localization in the history

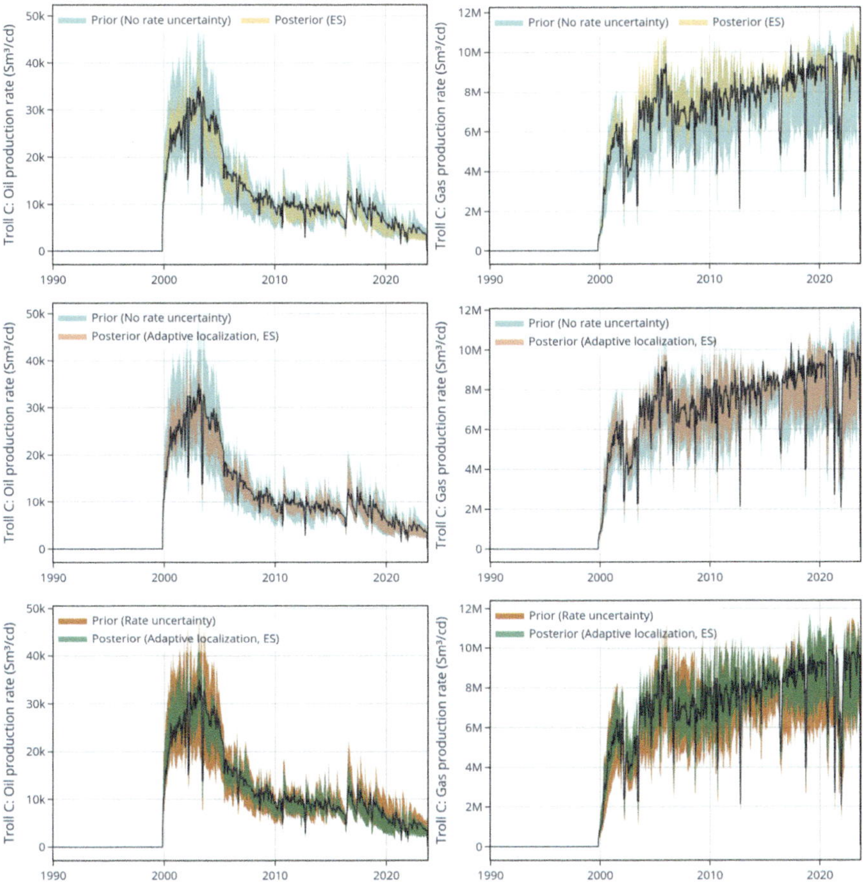

Fig. 14.9 The left and right plots show, respectively, the oil and gas production rates for Troll C for the global analysis (top), using adaptive localization but certain rates (second row), and using adaptive localization with uncertain rates (bottom)

matching. We observe that the results, when including the rate-control uncertainties and using adaptive localization, obtained the closest fit to the measured data, and the spread of the prediction forecasts is much higher than all the other cases.

14.7 Summary

This chapter discussed ensemble history matching with the reservoir model of the vast Troll field on the Norwegian continental shelf. The results demonstrate that ensemble methods are now generally feasible for use in reservoir management workflows. However, we must take care to ensure consistent use of the methods regarding the

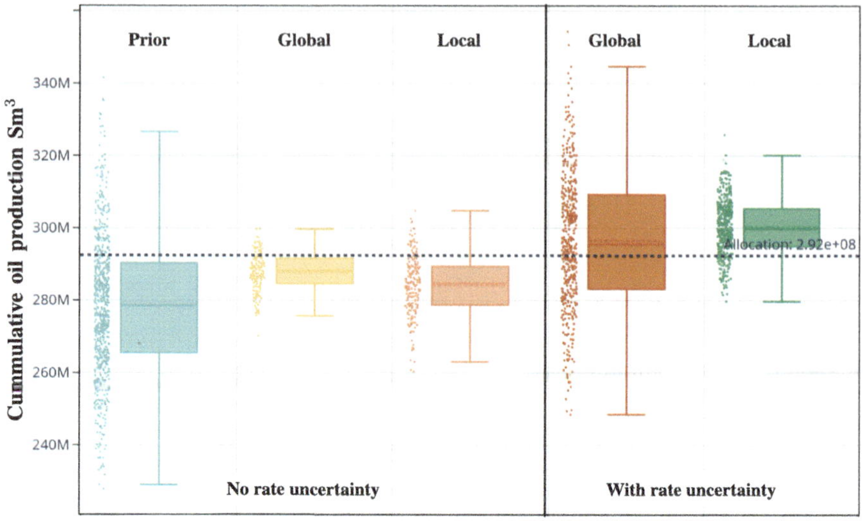

Fig. 14.10 Misfit: Troll West cumulative oil production at Sep 1, 2023. The dotted horizontal line is the allocation of 2.92e+08 Sm^3

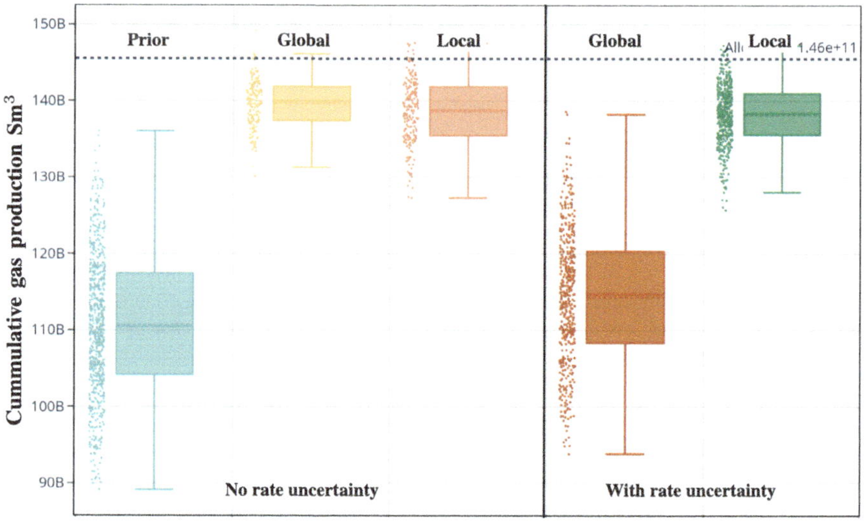

Fig. 14.11 Misfit: Troll B cumulative gas production at Sep 1, 2023. The dotted horizontal line is the allocation of 1.46e+11 Sm^3

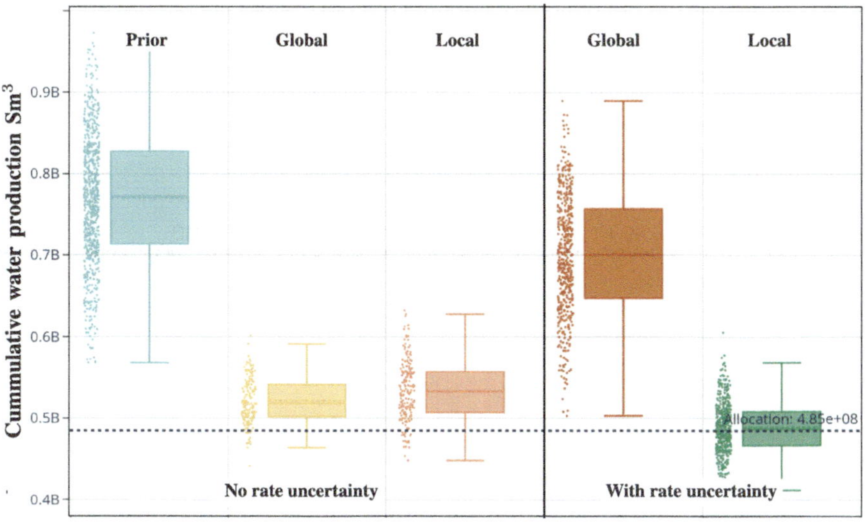

Fig. 14.12 Misfit: Troll C cumulative water production at Sep 1, 2023. The dotted horizontal line is the allocation of 4.85e+08 Sm^3

ensemble size, data preprocessing, and the specification of error statistics. An interesting aspect of the Troll example is the definition of a complete fast model update (FMU) model workflow involving the geological model, allowing us to history match geological parameters rather than update the simulation model parameters. The FMU approach is possible because the ensemble methods consider the model operator a "black box" and only use the ensemble correlations between the uncertain input parameters and the predicted measurements. Thus, FMU ensures consistency in the model chain and even allows updating structural parameters, leading to simulation models with different grids. In the final summary chapter, we will provide some thoughts about the future of reservoir history matching and closed-loop reservoir management.

Chapter 15
Summary and Future Perspectives

Abstract This final summary chapter discusses today's state of the art in ensemble history matching and robust optimization and gives a perspective on the future development of these methods and their potential for use in an operational closed-loop reservoir management systems.

15.1 State of the Art

The use of iterative ensemble smoothers is now well-established for history matching in the reservoir community, and the methods have proven their capabilities in several petroleum applications (Emerick and Reynolds 2012; Chen and Oliver 2014; Emerick 2016; Lorentzen et al. 2019; Evensen 2021; Oliver et al. 2021). The advantages of ensemble methods for history matching include the built-in uncertainty quantification, the possibility of defining high dimensional state vectors, and ease of implementation without the need for adjoints. The ensemble method's direct derivation from Bayes' theorem with clearly defined assumptions and approximations ensures mathematical rigor and consistency, which are essential to any operational method. The ensemble approach's drawbacks include the appearance of *sampling errors* caused by using a finite ensemble size, which introduces the need for localization and possibly inflation methods. Furthermore, the *Gaussianity* and *linearity* assumptions in an update step limit the level of nonlinearity in the reservoir model that the ensemble methods can handle, and much of the ensemble history-matching research focuses on resolving these three issues.

In this book, Chap. 7 discusses adaptive correlation-based localization methods as a remedy for sampling errors. Chapter 8 introduces the truncated and adaptive pluri-Gaussian methods to better handle categorical facies variables. An extensive discussion considers the extension from the linear ES update to the iterative EnRML or multi-step ESMDA methods to resolve the history-matching problem's general nonlinearity properties. On the application side, Chap. 13 discusses how to consistently configure and solve the ensemble history-matching problem and avoid common traps often seen in previous applications. Section 13.6 particularly emphasizes the value of understanding the impacts of conditioning models on redundant and

G. Evensen et al., *Ensemble History Matching*,
https://doi.org/10.1007/978-3-031-99155-4_15

dependent observations, a topic of utmost importance when consistently setting up a history-matching experiment and interpreting its results.

Robust optimization is a rapidly developing and recent research area partly motivated by the results from Van Essen et al. (2009), which demonstrated the value of optimizing controls over an ensemble of realizations representing the reservoir uncertainty. Ensemble Optimization (EnOpt), discussed in Chap. 10, is a popular robust optimization method that uses a stochastic gradient (Lorentzen et al. 2006; Chen et al. 2009), perfectly fitting the ensemble framework. Still, there is a need to make EnOpt more efficient as the stochastic gradient computed using models from an ensemble of geological models becomes noisy, and the convergence is slow, mainly when the ensemble spread is large (Chang and Evensen 2022). Chapter 11 discusses alternative methods that *optimize using the mean model* instead of the entire model ensemble. We have shown that such approximate methods may provide a near-optimal solution at a much lower cost. We expect significant further development of optimization methods suitable for robust reservoir optimization problems over the next few years.

Chapter 12 demonstrates how it is convenient to couple ensemble history matching with robust optimization methods to form *closed-loop reservoir management (CLRM) systems*. When using ensemble methods, the output of the history matching is a posterior ensemble of geological models conditioned on all historical dynamic data. This ensemble of conditioned models is input to the robust optimization, which leads to a vector of optimized controls for the future reservoir production or drilling schedule. These optimal controls result in the highest expected net present value (NPV) and sometimes the lowest expected emissions, depending on how we define our objective function. In the following sections, we will briefly discuss how it is possible to develop CLRMs with extensive automated reservoir modeling and simulation workflows, and we refer to some open-source workflow managers that should be useful for CLRM research and applications.

15.2 Automated Modeling Workflows

Traditionally, reservoir engineers have used history-matching methods to update a reservoir simulation model, which soon becomes incompatible with the underlying geological reservoir description represented by the geological model. However, when using ensemble methods, it is possible to do better. As illustrated by Fig. 1.4, when ensemble methods compute an update, they use the correlations between the input parameters and the predicted measurements to determine the magnitude of the update, and there is no need to know anything about the functional form of the model that maps the input parameters to the predicted measurements. The application with the Troll field in Chap. 14 takes advantage of this property of the ensemble methods. It defines a more advanced model workflow that includes the geological model in the forward operator. Thus, the workflow defines state variables that include Gaussian random fields as input parameters to the facies modeling, followed by running the simulation model with the updated facies realizations. In this case, it is straightforward to introduce a change of geological parameterization and automatically run the workflow through the geological modeling followed by running the simulation model.

The workflow can be even more general in other applications than in the Troll case. For example, one can simulate random structural surfaces and define the uncertain top and bottom reservoir surfaces. The challenge is to automate the structural modeling workflow to generate a new geological model grid from the input structural surfaces, populate the new grid with petrophysical properties, correctly introduce the wells in the new model grid, and then generate the new simulation model. Fortunately, some commercial modeling software makes it possible to automate or script this geological modeling process, leading to the Fast Model Update™ (FMU) workflow developed by Equinor (Hanea et al. 2015; Skjervheim et al. 2015) and described in Sect. 14.3.

The automated modeling workflows allow updating of geological parameters where they should be updated, i.e., in the geological model, in contrast to updating them in the simulation model. This strategy ensures consistency between the geological and simulation models and allows for a fast and straightforward updating of the geological model when new information becomes available. There is now a need for further research and development of automated modeling workflows and their use in ensemble history-matching.

15.3 Ensemble Workflow Manager

Although there are commercial software packages supporting ensemble workflows with history-matching and optimization functionalities, we will mention some alternatives that are publicly available for download. First, we will mention the Open Porous Media https://opm-project.org/ initiative that develops the Flow reservoir simulator for three-phase black-oil simulations. Flow includes support for CO_2 storage and thermal simulation and provides extensions to the black-oil model, such as polymer, solvent, or gas/water mixing. It can read and write standard industry formats. Being a high-quality open reservoir simulator, OPM Flow is well suited for ensemble integrations since there are no license requirements potentially limiting the number of concurrent parallel simulations. We used OPM Flow to run all simulations in Chap. 13.

The Python-Ensemble-Toolbox (PET) available from https://github.com/Python-Ensemble-Toolbox/PET is a toolbox for ensemble-based history matching and optimization, developed and maintained by the NORCE Research data assimilation and optimization group. The software is ready for download and is suitable for students and practitioners working with ensemble history matching and robust optimization.

The Ensemble Reservoir Tool (ERT) available from https://github.com/equinor/ert constitutes Equinor's operational in-house workflow manager for ensemble integrations, ensemble history matching, and optimization. It is software designed to run ensembles of dynamical models, such as reservoir models, for sensitivity analysis and data assimilation. ERT is also the software used for all Chaps. 13 and 14 experiments.

ERT and PET supports data assimilation using the Ensemble Smoother (ES), Ensemble Smoother with Multiple Data Assimilation (ESMDA), and the subspace Ensemble Randomized Likelihood (EnRML), implemented according to the algorithms developed in Chaps. 6 and with localization from Chap. 7. Thus, downloading a workflow manager and the OPM Flow reservoir simulator gives one all the essential tools to use ensemble methods.

15.4 Artificial Intelligence and Machine Learning

The petroleum industry is paying significant attention to incorporating methods for artificial intelligence (AI) and machine learning (ML) in reservoir management. Traditional reservoir simulation and optimization methods face increasing challenges due to computational limitations and the size of the involved data sets. Thus, AI and ML are promising approaches for more efficient reservoir modeling and streamlined decision-making processes.

While we do not explicitly discuss the adaptation of ML methods in this monograph, we recognize their potential within the petroleum industry. One of the most immediate applications of machine learning is in developing proxy models that mimic the behavior of more complex, computationally expensive reservoir simulations. These proxy models act as simplified versions of full-scale ensemble simulations, offering the ability to predict outcomes in a fraction of the time while maintaining high accuracy. We can then apply the ML methods to make faster, informed decisions about reservoir management and optimization strategies . Proxy models are beneficial in robust optimization scenarios, where multiple simulations account for uncertainty, improve the reliability of forecasts, and reduce the time and resources required for ensemble simulations.

Moreover, machine learning algorithms and reservoir history-matching methods share mathematical similarities. Both estimate models based on observed data to improve predictive accuracy. By combining the strengths of ensemble methods with the data-driven capabilities of machine learning, it seems likely that we will soon see a seamless integration of these methodologies.

Developing new hybrid approaches integrating machine learning and ensemble history matching will likely lead to faster reservoir optimization methods with higher accuracy. The petroleum industry expects a surge in the adoption of AI and ML-based solutions in the coming years, with these technologies playing a crucial role in addressing the industry's evolving challenges. In summary, integrating AI and ML methods with ensemble history matching and robust optimization has significant potential benefits. Thus, we expect the further development of these hybrid methods to redefine reservoir management, making processes more efficient and cost-effective.

15.5 Industry Adaptation Challenges

The introduction of ensemble modeling for uncertainty quantification and history matching in operational use in petroleum companies has been a long and tedious process. There are several reasons for the slow technology uptake.

Traditionally, reservoir engineers have not received training in using ensemble methods, and they usually work with a single model realization that they manually fine-tune to fit the dynamic data. When working with an ensemble of models, it is impossible to quality-check individual realizations, and one must work with the algorithms generating the model realizations instead. Thus, there is a significant change in the work processes and new competencies and mathematical understanding are required. As human beings, we are also often restrictive about change and prefer to continue as we are used to.

Clear leadership and management decisions are strongly needed to implement significant changes in a company, and first, the management needs to be convinced.

Another significant issue is the availability of software suitable for working with ensembles of reservoir models. Companies typically lock into specific vendors delivering software not originally designed to work with ensemble methods. Thus, there has been a long maturation and transition period where the vendors have started to support ensemble methods and have made new tools available to the industry.

Over the previous twenty years, methods and their applications have matured continuously, driven by the research community and the industry. Many presentations at petroleum conferences have demonstrated the benefits of using ensemble methods, which are generally accepted as state-of-the-art today.

Regarding the use of interdisciplinary modeling workflows like FMU, the technical staff in the companies are often working "almost independently" within different disciplines. For example, a geophysicist may deliver a seismic interpretation to the geologist who builds a detailed geological model. The geologist hands the model over to the reservoir engineer, who implements the reservoir simulation model. When introducing an FMU workflow in a company, the geological modeler must change their way of working to develop the workflow or scripts that generate the geological model, rather than manually building and maintaining a geological model. Furthermore, a reservoir engineer will update the input parameters to the geological model, and this cross-disciplinary work requires direct hands-on collaboration between the different disciplines.

References

Aanonsen, S. I., G. Naevdal, D. S. Oliver, A. C. Reynolds, and B. Valles. 2009. Ensemble Kalman filter in reservoir engineering–A review. *SPE Journal* 14 (3): 393–412. https://doi.org/10.2118/117274-PA.

Abadpour, A., M. Adejare, T. Chugunova, H. Matheo, and N. Haller. 2018. Integrated geo-modeling and ensemble history matching of complex fractured carbonate and deep offshore turbidite fields. *SPE Conference Paper*. https://doi.org/10.2118/193028-MS.

Agbalaka, C., and D. Oliver. 2009. Automatic history matching of production and facies data with nonstationary proportions using EnKF. *SPE Reservoir Simulation Symposium Proceedings* 1: 02. https://doi.org/10.2118/118916-MS.

Agbalaka, C. C., and D. S. Oliver. 2008. Application of the EnKF and localization to automatic history matching of facies distribution and production data. *Mathematical Geosciences* 40 (4): 353–374. https://doi.org/10.1007/s11004-008-9155-7.

Aghajani, D., R. B. Bratvold, V. Hagspiel, O. Noshchenko, and V. K. Toutain. 2025. A multi-objective decision-making framework for the choice between mutually exclusive alternatives under uncertainty: Assessing the competitiveness of offshore wind for a gas field electrification on the NCS. *Energy Economics* 141: 108032. https://doi.org/10.1016/j.eneco.2024.108032.

Ahmadi, R., and R. B. Bratvold. 2023. An exposition of least square monte carlo approach for real options valuation. *Geoenergy Science and Engineering* 222: 111230. https://doi.org/10.1016/j.petrol.2022.111230.

Anderson, J. L. 2003. A local least squares framework for ensemble filtering. *Monthly Weather Review* 131: 634–642.

Anderson, J. L. 2007. An adaptive covariance inflation error correction algorithm for ensemble filters. *Tellus, Series A* 59: 210–224. https://doi.org/10.1111/j.1600-0870.2006.00216.x.

Anderson, J. L. 2009. Spatially and temporally varying adaptive covariance inflation for ensemble filters. *Tellus, Series A* 61: 72–83. https://doi.org/10.1111/j.1600-0870.2008.00361.x.

Anderson, J. L. 2012. Localization and sampling error correction in ensemble Kalman filter data assimilation. *Monthly Weather Review* 140: 2359–2371. DOI: https://doi.org/10.1175/MWR-D-11-00013.1.

Anderson, J. L. 2016. Reducing correlation sampling error in ensemble Kalman filter data assimilation. *Monthly Weather Review* 144 (3): 913–925. https://doi.org/10.1175/MWR-D-15-0052.1.

Armstrong, M., A. Galli, H. Beucher, G. Le Loc'h, D. Renard, B. Doligez, R. Eschard, and F. Geffroy. 2011. *Plurigaussian Simulations in Geosciences, volume*, 2nd ed. Berlin, Heidelberg: Springer. https://doi.org/10.1007/978-3-642-19607-2.

G. Evensen et al., *Ensemble History Matching*,
https://doi.org/10.1007/978-3-031-99155-4

Astrakova, A., and D. S. Oliver. 2014. Conditioning truncated pluri-Gaussian models to facies observations in ensemble-Kalman-based data assimilation. *Mathematical Geosciences* 47 (3): 345–367. https://doi.org/10.1007/s11004-014-9532-3.

Ba, Y., J. Wiljes, D. S. Oliver, and S. Reich. 2022. Randomized maximum likelihood based posterior sampling. *Computational Geosciences* 26 (1): 217–239. https://doi.org/10.1007/s10596-021-10100-y.

Bangerth, W., H. Klie, M. F. Wheeler, P. L. Stoffa, and M. K. Sen. 2006. On optimization algorithms for the reservoir oil well placement problem. *Computational Geosciences* 10 (3): 303–319. https://doi.org/10.1007/s10596-006-9025-7.

Bennett, A. F. 1992. Inverse methods in physical oceanography. *Cambridge University Press*. https://doi.org/10.1017/CBO9780511600807.

Berahas, A. S., R. H. Byrd, and J. Nocedal. 2019. Derivative-free optimization of noisy functions via quasi-Newton methods. *SIAM Journal on Optimization* 29 (2): 965–993. https://doi.org/10.1137/18M1177718.

Berahas, A. S., L. Cao, K. Choromanski, and K. Scheinberg. 2022. A theoretical and empirical comparison of gradient approximations in derivative-free optimization. *Foundations of Computational Mathematics* 22 (2): 507–560. https://doi.org/10.1007/s10208-021-09513-z.

Bianco, A., A. Cominelli, L. Dovera, G. Naevdal, and B. Valles. 2007. History matching and production forecast uncertainty by means of the ensemble Kalman filter: A real field application. *SPE Conference Paper*. https://doi.org/10.2118/107161-MS.

Bishop, C. H., and D. Hodyss. 2007. Flow-adaptive moderation of spurious ensemble correlations and its use in ensemble-based data assimilation. *The Quarterly Journal of the Royal Meteorological Society* 133: 2029–2044. https://doi.org/10.1002/qj.169.

Bishop, C. H., and D. Hodyss. 2009a. Ensemble covariances adaptively localized with ECO-RAP. Part 1: Tests on simple error models. *Tellus, Series A*, 61A: 84–96. https://onlinelibrary.wiley.com/doi/10.1111/j.1600-0870.2008.00371.x.

Bishop, C. H., and D. Hodyss. 2009b. Ensemble covariances adaptively localized with ECO-RAP. Part 2: A strategy for the atmosphere. *Tellus, Series A*, 61A: 97–111. https://onlinelibrary.wiley.com/doi/10.1111/j.1600-0870.2007.00372.x.

Bishop, C. H., B. J. Etherton, and S. J. Majumdar. 2001. Adaptive sampling with the ensemble transform Kalman filter. Part I: Theoretical aspects. *Monthly Weather Review* 129: 420–436. https://doi.org/10.1175/1520-0493(2001)129<0420:ASWTET>2.0.CO;2.

Bissell, R., Y. Sharma, and J. E. Killough. 1994. History matching using the method of gradients: Two case studies. SPE 69th Annual Technical Conference and Exhibition. *SPE* 28590: 275–289. https://doi.org/10.2118/28590-MS.

Bocquet, M., P. N. Raanes, and A. Hannart. 2015. Expanding the validity of the ensemble Kalman filter without the intrinsic need for inflation. *Nonlinear Processes in Geophysics* 22 (6): 645–662. https://doi.org/10.5194/npg-22-645-2015.

Box, G. E. P., and D. R. Cox. 1964. An analysis of transformations. *Journal of Royal Statistical Society, Series B* 26: 211–243. https://doi.org/10.1111/j.2517-6161.1964.tb00553.x.

Bratvold, R. B., J. E. Bickel, and H. P. Lohne. 2009. Value of information in the oil and gas industry: Past, present, and future. *SPE Reservoir Evaluation and Engineering*, 12(04): 630–638, 09 2009. https://doi.org/10.2118/110378-PA.

Brouwer, D. R., G. Nævdal, J. D. Jansen, E. H. Vefring, and C. P. J. W. van Kruijsdijk. 2004. Improved reservoir management through optimal control and continuous model updating. In *SPE annual technical conference and exhibition*, Sept 26–29, Houston, Texas. https://doi.org/10.2118/90149-MS.

Burgers, G., P. J. Leeuwen, and G. Evensen. 1998. Analysis scheme in the ensemble Kalman filter. *Monthly Weather Review* 126: 1719–1724. https://doi.org/10.1175/1520-0493(1998)126<1719:ASITEK>2.0.CO;2.

Caers, J., and T. Zhang. 2004. Multiple-point geostatistics: A quantitative vehicle for integrating geologic analogs into multiple reservoir models. In *Integration of outcrop and modern analogs*

in reservoir modeling. American Association of Petroleum Geologists. https://doi.org/10.1306/M80924C18.
Carrassi, A., M. Bocquet, L. Bertino, and G. Evensen. 2018. Data assimilation in the geosciences: An overview on methods, issues and perspectives. *Wires Climate Change* 9 (5): 50. https://doi.org/10.1002/wcc.535.
Cartis, C., J. Fiala, B. Marteau, and L. Roberts. 2019. Improving the flexibility and robustness of model-based derivative-free optimization solvers. *ACM Transactions on Mathematical Software*, 45(3). https://doi.org/10.1145/3338517.
Casella, G., and R. L. Berger. 2024. *Statistical inference*, 2nd ed. Chapman and Hall/CRC. https://doi.org/10.1201/9781003456285.
Chang, Y., and G. Evensen. 2022. An ensemble-based decision workflow for reservoir management. *The Journal of Petroleum Science and Engineering*. https://doi.org/10.1016/j.petrol.2022.110858.
Chang, Y., R. J. Lorentzen, G. Nævdal, and T. Feng. 2020. Olympus optimization under geological uncertainty. *Computers and Geosciences* 24: 2027–2042. https://doi.org/10.1007/s10596-019-09892-x.
Chen, C., Y. Wang, G. Li, and A. C. Reynolds. 2010. Closed-loop reservoir management on the Brugge test case. *Computational Geosciences* 14 (4): 691–703. https://doi.org/10.1007/s10596-010-9181-7.
Chen, Y., and D. S. Oliver. 2010. Ensemble-based closed-loop optimization applied to Brugge Field. *SPE Reservoir Evaluation and Engineering* 13 (1): 56–71. https://doi.org/10.2118/118926-PA.
Chen, Y., and D. S. Oliver. 2010. Cross-covariances and localization for EnKF in multiphase flow data assimilation. *Computational Geosciences* 14: 579–601. https://doi.org/10.1007/s10596-009-9174-6.
Chen, Y., and D. S. Oliver. 2012. Ensemble randomized maximum likelihood method as an iterative ensemble smoother. *Mathematical Geosciences* 44: 1–26. https://doi.org/10.1007/s11004-011-9376-z.
Chen, Y., and D. S. Oliver. 2013. Levenberg-Marquardt forms of the iterative ensemble smoother for efficient history matching and uncertainty quantification. *Computational Geosciences* 17: 689–703. https://doi.org/10.1007/s10596-013-9351-5.
Chen, Y., and D. S. Oliver. 2014. History matching of the Norne full-field model using an iterative ensemble smoother. *SPE Reservoir Evaluation and Engineering* 17 (2): 244–256. https://doi.org/10.2118/164902-PA.
Chen, Y., and D. S. Oliver. 2017. Localization and regularization for iterative ensemble smoothers. *Computational Geosciences* 21: 13–30. https://doi.org/10.1007/s10596-016-9599-7.
Chen, Y., D. S. Oliver, and D. Zhang. 2009. Efficient ensemble-based closed-loop production optimization. *SPE Journal* 14 (4): 634–645. https://doi.org/10.2118/112873-PA.
Cominelli, A., L. Dovera, S. Vimercati, and G. Nævdal. 2009. Benchmark study of ensemble Kalman filter methodology: History matching and uncertainty quantification for a deep-water oil reservoir. In *IPTC International Petroleum Technology Conference*, IPTC–13748–MS. https://doi.org/10.2523/IPTC-13748-MS.
Conn, A. R., K. Scheinberg, and L. N. Vicente. 2009. *Introduction to derivative-free optimization*. MPS-SIAM series on optimization: SIAM 978-0-898716-68-9. https://doi.org/10.1137/1.9780898718768.
Cosentino, L. 2001. *Integrated Reservoir Studies*. Editions Technips. ISBN 978-27108-0797-1. https://www.editionstechnip.com/en/catalogue-detail/695/integrated-reservoir-studies.html.
Denney, D. 2008. Troll West oil field–A giant gas field becomes the largest oil field on the Norwegian Continental Shelf. *Journal of Petroleum Technology* 60 (10): 47–49. https://doi.org/10.2118/1008-0047-JPT.
Deraisme, J., and D. Farrow. 2005. Geostatistical simulation techniques applied to kimberlite orebodies and risk assessment of sampling strategies, 429–438. *Springer, Netherlands*. https://doi.org/10.1007/978-1-4020-3610-1_43.

Dreyer, T., M. Whitaker, J. Dexter, H. Flesche, and E. Larsen. 2005. From spit system to tide-dominated delta: Integrated reservoir model of the Upper Jurassic Sognefjord Formation on the Troll West Field. *Geological Society, London, Petroleum Geology Conference series* 6. DOI: https://doi.org/10.1144/0060423.

Eiken, O., T. Stenvold, M. Zumberge, H. Alnes, and G. Sasagawa. 2008. Gravimetric monitoring of gas production from the Troll field. *Geophysics* 73: 11. https://doi.org/10.1190/1.2978166.

Emerick, A., and A. C. Reynolds. 2011. Combining sensitivities and prior information for covariance localization in the ensemble Kalman filter for petroleum reservoir applications. *Computational Geosciences* 15 (2): 251–269. https://doi.org/10.1007/s10596-010-9198-y.

Emerick, A. A. 2016. Analysis of performance of ensemble-based assimilation of production and seismic data. *Journal of Petroleum Science and Engineering* 139: 219–239. https://doi.org/10.1016/j.petrol.2016.01.029.

Emerick, A. A. 2018. Deterministic ensemble smoother with multiple data assimilation as an alternative for history matching seismic data. *Computational Geosciences* 22: 10. https://doi.org/10.1007/s10596-018-9745-5.

Emerick, A. A. 2019. Analysis of geometric selection of the data-error covariance inflation for ES-MDA. *Journal of Petroleum Science and Engineering* 182: 106168. https://doi.org/10.1016/j.petrol.2019.06.032.

Emerick, A. A., and G. M. S. Neto. 2023. Investigation on the production data frequency for assimilation with ensemble smoother. *Geoenergy Science and Engineering* 231. DOI: https://doi.org/10.1016/j.geoen.2023.212356.

Emerick, A. A., and A. C. Reynolds. 2012. History matching time-lapse seismic data using the ensemble Kalman filter with multiple data assimilations. *Computational Geosciences* 16 (3): 639–659. https://doi.org/10.1007/S10596-012-9275-5.

Emerick, A. A., and A. C. Reynolds. 2013. Ensemble smoother with multiple data assimilation. *Computers and Geosciences* 55: 3–15. https://doi.org/10.1016/j.cageo.2012.03.011.

Emery, X. 2007. Simulation of geological domains using the plurigaussian model: New developments and computer programs. *Computers and Geosciences* 33 (9): 1189–1201. https://doi.org/10.1016/j.cageo.2007.01.006.

Emery, X. 2008. Statistical tests for validating geostatistical simulation algorithms. *Computers and Geosciences* 34 (11): 1610–1620. https://doi.org/10.1016/j.cageo.2007.12.012.

Evensen, G. 1994. Sequential data assimilation with a nonlinear quasi-geostrophic model using Monte Carlo methods to forecast error statistics. *Journal of Geophysical Research* 99 (C5): 10143–10162. https://doi.org/10.1029/94JC00572.

Evensen, G. 2003. The ensemble Kalman filter: Theoretical formulation and practical implementation. *Ocean Dynamics* 53: 343–367. https://doi.org/10.1007/s10236-003-0036-9.

Evensen, G. 2004. Sampling strategies and square root analysis schemes for the EnKF. *Ocean Dynamics* 54: 539–560. https://doi.org/10.1007/s10236-004-0099-2.

Evensen, G. 2007. *Data Assimilation: The Ensemble Kalman Filter*, 1st ed. Springer. https://doi.org/10.1007/978-3-642-03711-5.

Evensen, G. 2009. *Data Assimilation: The Ensemble Kalman Filter*, 2nd ed. Springer. https://doi.org/10.1007/978-3-642-03711-5.

Evensen, G. 2018. Analysis of iterative ensemble smoothers for solving inverse problems. *Computational Geosciences* 22 (3): 885–908. https://doi.org/10.1007/s10596-018-9731-y.

Evensen, G. 2019. Accounting for model errors in iterative ensemble smoothers. *Computational Geosciences* 23 (4): 761–775. https://doi.org/10.1007/s10596-019-9819-z.

Evensen, G. 2021. Formulating the history matching problem with consistent error statistics. *Computational Geosciences* 25: 945–970. https://doi.org/10.1007/s10596-021-10032-7.

Evensen, G., and Y. Chang. 2022. An ensemble-based decision-making workflow for reservoir management. In *Proceedings of the European conference on the mathematics of oil recovery (ECMOR 2022)*, The Hauge, The Netherlands, Sept 5–7, 2022. https://doi.org/10.3997/2214-4609.202244061.

Evensen, G., and K. S. Eikrem. 2018. Strategies for conditioning reservoir models on rate data using ensemble smoothers. *Computational Geosciences* 22 (5): 1251–1270. https://doi.org/10.1007/s10596-018-9750-8.

Evensen, G., and P. J. Leeuwen. 2000. An ensemble Kalman smoother for nonlinear dynamics. *Monthly Weather Review* 128: 1852–1867. https://doi.org/10.1175/1520-0493(2000)128<1852:AEKSFN>2.0.CO;2.

Evensen, G., J. Hove, H. C. Meisingset, E. Reiso, K. S. Seim, and Ø. Espelid. 2007. Using the EnKF for assisted history matching of a North Sea reservoir model. *SPE Conference Paper*. https://doi.org/10.2118/106184-MS.

Evensen, G., P. N. Raanes, A. S. Stordal, and J. Hove. 2019. Efficient implementation of an iterative ensemble smoother for data assimilation and reservoir history matching. *Frontiers in Applied Mathematics and Statistics* 5: 47. https://doi.org/10.3389/fams.2019.00047.

Evensen, G., F. C. Vossepoel, and P. J. Van Leeuwen. Data Assimilation Fundamentals: A Unified formulation for State and Parameter Estimation. Springer. 2022. ISBN 978-3-030-96708-6. *Open Access*. https://doi.org/10.1007/978-3-030-96709-3.

Evensen, G., F. C. Vossepoel, and P. J. Van Leeuwen. 2024. Iterative ensemble smoothers for data assimilation in coupled nonlinear multiscale models. *Monthly Weather Review* 152. DOI: https://doi.org/10.1175/MWR-D-23-0239.1.

Flowerdew, J. 2015. Towards a theory of optimal localisation. *Tellus, Series A* 67. DOI: https://doi.org/10.3402/tellusa.v67.25257.

Fonseca, R. 2016. *A modified gradient formulation for ensemble optimization under geological uncertainty*. Ph.D. thesis, Delft University of Technology. http://resolver.tudelft.nl/uuid:e66b1e00-b4c2-43b8-91fa-c57773fcf24b.

Galli, A., H. Beucher, G. Le Loc'h, B. Doligez, and H. Group. 1994. The pros and cons of the truncated Gaussian method. In Armstrong, M., and P. A. Dowd, editors, *Geostatistical simulations*, 217–233, Dordrecht. Springer Netherlands. https://doi.org/10.1007/978-94-015-8267-4_18.

Gamba, A. 2003. Real options valuation: a Monte Carlo approach. In *Faculty of Management, University of Calgary WP No. 2002/3; EFA 2002 Berlin Meetings Presented Paper WBS Finance Group Research Paper No. 14*.https://doi.org/10.2139/ssrn.302613.

Griewank, A. 2003. A mathematical view of automatic differentiation. *Acta Numerica* 12: 321–398. https://doi.org/10.1017/S0962492902000132.

Gu, Y., and D. S. Oliver. 2005. History matching of the PUNQ-S3 reservoir model using the ensemble Kalman filter. *SPE Journal* 10 (2): 217–224. https://doi.org/10.2118/89942-PA.

Gu, Y., and D. S. Oliver. 2007. An iterative ensemble Kalman filter for multiphase fluid flow data assimilation. *SPE Journal* 12 (4): 438–446. https://doi.org/10.2118/108438-PA.

Hamill, T. M. 2001. Interpretation of rank histograms for verifying ensemble forecasts. *Monthly Weather Review* 129: 550–560. https://doi.org/10.1175/1520-0493(2001)129<0550:IORHFV>2.0.CO;2.

Hanea, R., G. Evensen, L. Hustoft, T. Ek, A. Chitu, and F. Wilschut. 2015. Reservoir management under geological uncertainty using Fast Model Update. *SPE conference paper*. https://onepetro.org/spersc/proceedings-abstract/15RSS/15RSS/D031S009R003/182343.

Hanea, R. G., P. Casanova, F. Wilschut, L. Hustoft, and R. M. Fonseca. 2017. Well trajectory optimization constrained to structural uncertainties. *SPE conference paper* 02: 2017. https://doi.org/10.2118/182680-MS.

Haugen, V. E., O. M. Johannessen, and G. Evensen. 2002. Indian ocean: validation of the Miami Isopycnic Coordinate ocean model and ENSO events during 1958–1998. *Journal of Geophysical Research*, 107(C5): 11–1–11–23. https://doi.org/10.1007/s10236-002-0014-7.

Haugen, V. E., G. Nævdal, L.-J. Natvik, G. Evensen, A. Berg, and K. Flornes. 2008. History matching using the ensemble Kalman filter on a North Sea field case. *SPE Journal*, 13: 382–391, 12 2008. https://doi.org/10.2118/102430-PA.

Houtekamer, P. L., and H. L. Mitchell. 1998. Data assimilation using an Ensemble Kalman Filter technique. *Monthly Weather Review* 126: 796–811. https://doi.org/10.1175/1520-0493(1998)126<DAUAEK>2.0.CO;2.

Houtekamer, P. L., and H. L. Mitchell. 2001. A sequential ensemble Kalman filter for atmospheric data assimilation. *Monthly Weather Review* 129: 123–137. https://doi.org/10.1175/1520-0493(2001)129<ASEKFF>2.0.CO;2.

Houtekamer, P. L., and F. Zhang. 2016. Review of the ensemble Kalman filter for atmospheric data assimilation. *Monthly Weather Review* 144: 4489–4533. https://doi.org/10.1175/MWR-D-15-0440.1.

Howard, R. 1960. *Dynamic Programming and Markov Processes*. Technology Press of Massachusetts Institute of Technology: Technology Press Research Monographs.

Howard, R., and A. Abbas. 2015. *Foundations of Decision Analysis*. Pearson Education, 1st ed. ISBN 978-0-133-48595-0. https://www.pearson.com/en-us/subject-catalog/p/foundations-of-decision-analysis/P200000003532/9780132336246.

Hunt, B. R., E. J. Kostelich, and I. Szunyogh. 2007. Efficient data assimilation for spatiotemporal chaos: A local ensemble transform Kalman filter. *Physica D* 230: 112–126. https://doi.org/10.1016/j.physd.2006.11.008.

Jafarpour, B., and D. B. McLaughlin. 2007. History matching with an ensemble Kalman filter and discrete cosine parameterization. *SPE Conference Paper*. https://doi.org/10.2118/108761-MS.

Jansen, J. D. 2011. Adjoint-based optimization of multi-phase flow through porous media–A review. *Computers and Fluids* 46 (1): 40–51. https://doi.org/10.1016/j.compfluid.2010.09.039.

Kitanidis, P. K. 1995. Quasi-linear geostatistical theory for inversing. *Water Resources Research* 31 (10): 2411–2419. https://doi.org/10.1029/95WR01945.

Lantuéjoul, C. 2002. Geostatistical Simulation: Models and Algorithms. *Springer*. https://doi.org/10.1007/978-3-662-04808-5.

Larson, J., M. Menickelly, and S. M. Wild. 2019. Derivative-free optimization methods. *Acta Numerica* 28: 287–404. https://doi.org/10.1017/S0962492919000060.

Le Loc'h, G. and A. Galli. 1997. Truncated pluri-Gaussian method: Theoretical and practical points of view. In Baafi, E. Y., and N. A. Schofield, editors, *Geostatistics Wollongong' 96*, vol. 1, 211–222. Springer Dordrecht. ISBN 978-0-7923-4496-4. https://link.springer.com/book/9780792344964.

Le Loc'h, G., H. Beucher, A. Galli, and B. Doligez. 1994. Improvement in the truncated Gaussian method: Combining several Gaussian functions. In *4th European conference on the mathematics of oil recovery, Røros, Norway*, 13. European Association of Geoscientists and Engineers. https://doi.org/10.3997/2214-4609.201411149.

Le Ravalec-Dupin, M., F. Roggero, and R. Froidevaux. Conditioning truncated Gaussian realizations to static and dynamic data. *SPE Journal*, 9(04): 475–480, 12 2004. https://doi.org/10.2118/84944-PA.

Leeuwenburgh, O., A. Chitu, R. Nair, P. Egberts, L. Ghazaryan, T. Feng, and L. Hustoft. 2016. Ensemble-based methods for well drilling sequence and time optimization under uncertainty. In *ECMOR XV-15th European conference on the mathematics of oil recovery*. European Association of Geoscientists and Engineers. https://doi.org/10.3997/2214-4609.201601871.

Li, G., and A. C. Reynolds. 2009. Iterative ensemble Kalman filters for data assimilation. *SPE Journal*, 14(03):496–505, 09 2009. https://doi.org/10.2118/109808-PA.

Li, R., A. C. Reynolds, and D. S. Oliver. 2003. Sensitivity coefficients for three-phase flow history matching. *Journal of Canadian Petroleum Technology* 42 (4): 70–77. https://doi.org/10.2118/03-04-04.

Liu, N., and D. S. Oliver. 2005. Ensemble Kalman filter for automatic history matching of geologic facies. *Journal of Petroleum Science and Engineering* 47 (3): 147–161. https://doi.org/10.1016/j.petrol.2005.03.006.

Longstaff, F. A., and E. S. Schwartz. 2015. Valuing American options by simulation: A simple least-squares approach. *The Review of Financial Studies* 14 (1): 113–147. https://doi.org/10.1093/rfs/14.1.113.

Lorentzen, R. J., K. K. Fjelde, J. Frøyen, A. C. V. M. Lage, G. Nævdal, and E. H. Vefring. 2001. Underbalanced and low-head drilling operations: Real time interpretation of measured data and

operational support. *SPE Annual Technical Conference and Exhibition*. https://doi.org/10.2118/71384-MS.
Lorentzen, R. J., G. Nævdal, and A. C. V. M. Lage. 2003. Tuning of parameters in a two-phase flow model using an ensemble Kalman filter. *International Journal of Multiphase Flow* 29: 1283–1309. https://doi.org/10.1016/S0301-9322(03)00088-0.
Lorentzen, R. J., A. M. Berg, G. Nævdal, and E. H. Vefring. 2006. A new approach for dynamic optimization of water flooding problems. *SPE Conference Paper*. https://doi.org/10.2118/99690-MS.
Lorentzen, R. J., X. Luo, T. Bhakta, and R. Valestrand. 2019. History matching the full Norne field model using seismic and production data. *SPE Journal* 24: 1452–1467. https://doi.org/10.2118/194205-PA.
Luo, X., and T. Bhakta. 2020. Automatic and adaptive localization for ensemble-based history matching. *Journal of Petroleum Science and Engineering*, 184: 106559. ISSN 0920-4105. https://doi.org/10.1016/j.petrol.2019.106559.
Luo, X., and C.-A. Xia. 2022. Continuous hyper-parameter optimization (CHOP) in an ensemble Kalman filter. *Frontiers in Applied Mathematics and Statistics* 8. DOI: https://doi.org/10.3389/fams.2022.1021551.
Madsen, T., and M. Abtahi. 2005. *Handling the oil zone on Troll*.
Matheron, G., H. Beucher, C. de Fouquet, A. Galli, D. Guerillot, and C. Ravenne. 1987. Conditional simulation of the geometry of fluvio-deltaic reservoirs. *SPE Annual Technical Conference and Exhibition*. https://doi.org/10.2118/16753-MS.
Mattax, C. C., and R. L. Dalton. 1990. Reservoir simulation (includes associated papers 21606 and 21620). *Journal of Petroleum Technology* 42 (06): 692–695. https://doi.org/10.2118/20399-PA.
Mirzaei-Paiaman, A., S. M. G. Santos, and D. J. Schiozer. 2021. A review on closed-loop field development and management. *Journal of Petroleum Science and Engineering* 201: 108457. https://doi.org/10.1016/j.petrol.2021.108457.
Mjaavatten, A., R. Aasheim, S. Saelid, and O. Groenning. 2008. A model for gas coning and rate-dependent gas/oil ratio in an oil-rim reservoir. *SPE Reservoir Evaluation and Engineering* 11: 10. https://doi.org/10.2118/102390-MS.
Moré, J. J., and S. M. Wild. 2009. Benchmarking derivative-free optimization algorithms. *SIAM Journal on Optimization* 20 (1): 172–191. https://doi.org/10.1137/080724083.
Nævdal, G., T. Mannseth, and E. Vefring. 2002. Near well reservoir monitoring through ensemble Kalman filter. *SPE Conference Paper*. https://doi.org/10.2118/75235-MS.
Nævdal, G., L. M. Johnsen, S. I. Aanonsen, and E. Vefring. 2005. Reservoir monitoring and continuous model updating using the ensemble Kalman filter. *SPE Journal* 10 (1): 66–74. https://doi.org/10.2118/84372-PA.
Nævdal, G., D. R. Brouwer, and J.-D. Jansen. 2006. Waterflooding using closed-loop control. *Computational Geosciences* 10 (1): 37–60. https://doi.org/10.1007/s10596-005-9010-6.
Neal, R. M. 1996. Sampling from multimodal distributions using tempered transitions. *Statistics and Computing* 6 (4): 353–366. https://doi.org/10.1007/BF00143556.
Neto, G. M. S., R. V. Soares, G. Evensen, A. Davolioa, and D. J. Schiozer. 2021. Subspace ensemble randomized maximum likelihood with local analysis for time-lapse-seismic-data assimilation. *SPE Journal* 26 (2): 1011–1031. https://doi.org/10.2118/205029-PA.
Oliver, D. S. 2024. Robust optimization using the mean model with bias correction. *Mathematical Geosciences*. https://doi.org/10.1007/s11004-024-10155-4.
Oliver, D. S., and Y. Chen. 2011. Recent progress on reservoir history matching: A review. *Computational Geosciences* 15 (1): 185–221. https://doi.org/10.1007/s10596-010-9194-2.
Oliver, D. S., N. He, and A. C. Reynolds. 1996. Conditioning permeability fields to pressure data. In *ECMOR V-5th European conference on the mathematics of oil recovery*, 11. https://doi.org/10.3997/2214-4609.201406884.
Oliver, D. S., A. C. Reynolds, and N. Liu. 2008. Inverse Theory for Petroleum Reservoir Characterization and History Matching. *Cambridge University Press, Cambridge, UK*. https://doi.org/10.1017/CBO9780511535642.

Oliver, D. S., K. Fossum, T. Bhakta, I. Sandø, G. Nævdal, and R. J. Lorentzen. 2021. 4D seismic history matching. *Journal of Petroleum Science and Engineering* 207. DOI: https://doi.org/10.1016/j.petrol.2021.109119.

Perrone, A., F. Pennadoro, A. Tiani, E. Della Rossa, and J. Saetrom. 2017. Enhancing the geological models consistency in ensemble based history matching an integrated approach. *Society of Petroleum Engineers*. https://doi.org/10.2118/186049-MS.

Peters, E., R. J. Arts, G. K. Brouwer, C. R. Geel, S. Cullick, R. J. Lorentzen, Y. Chen, K. N. B. Dunlop, F. C. Vossepoel, R. Xu, P. Sarma, A. H. Alhutali, and A. C. Reynolds. 2010. Results of the Brugge benchmark study for flooding optimization and history matching. *SPE Reservoir Evaluation and Engineering* 13: 391–405. https://doi.org/10.2118/119094-PA.

Powell, M. J. D. 2006. The NEWUOA software for unconstrained optimization without derivatives. In Pillo, G. and M. Roma, editors, *Large-Scale Nonlinear Optimization*, 255–297. Springer. https://doi.org/10.1007/0-387-30065-1_16.

Powell, M. J. D. 2009. The BOBYQA algorithm for bound constrained optimization without derivatives. https://api.semanticscholar.org/CorpusID:2488733.

Powell, W. B. 2011. *Approximate Dynamic Programming: Solving the Curses of Dimensionality*. Wiley Series in Probability and Statistics. Wiley. ISBN 9780470604458.https://doi.org/10.1002/9781118029176.

Raanes, P. N., A. S. Stordal, and G. Evensen. 2019. Revising the stochastic iterative ensemble smoother. *Nonlinear Processes in Geophysics* 26: 325–338. https://doi.org/10.5194/npg-2019-10.

Rafiee, J., and A. C. Reynolds. 2017. Theoretical and efficient practical procedures for the generation of inflation factors for ES-MDA. *Inverse Problems, online* 33. DOI: https://doi.org/10.1088/1361-6420/aa8cb2.

Ragonneau, T. M., and Z. Zhang. 2024. PDFO: A cross-platform package for Powell's derivative-free optimization solvers.

Sakov, P., D. S. Oliver, and L. Bertino. 2012. An iterative EnKF for strongly nonlinear systems. *Monthly Weather Review* 140: 1988–2004. https://doi.org/10.1175/MWR-D-11-00176.1.

Schulze-Riegert, R. W., J. K. Axmann, O. Haase, D. T. Rian, and Y. L. You. 2002. Evolutionary algorithms applied to history matching of complex reservoirs. *SPE Reservoir Evaluation and Engineering* 5 (2): 163–173. https://doi.org/10.2118/77301-PA.

Sebacher, B., R. Hanea, and A. Heemink. 2013. A probabilistic parametrization for geological uncertainty estimation using the ensemble Kalman filter (EnKF). *Computational Geosciences* 17 (5): 813–832. https://doi.org/10.1007/s10596-013-9357-z.

Sebacher, B., A. Stordal, and R. Hanea. 2015. Bridging multi-point statistics and truncated Gaussian fields for improved estimation of channelized reservoirs with ensemble methods. *Computational Geosciences* 19 (2): 341–369. https://doi.org/10.1007/s10596-014-9466-3.

Sebacher, B., A. Stordal, and R. Hanea. 2016. Complex geology estimation using the iterative adaptive Gaussian mixture filter. *Computational Geosciences* 20 (1): 133–148. https://doi.org/10.1007/s10596-015-9553-0.

Sebacher, B., R. Hanea, and A. S. Stordal. 2017. An adaptive pluri-Gaussian simulation model for geological uncertainty quantification. *Journal of Petroleum Science and Engineering* 158: 494–508. https://doi.org/10.1016/j.petrol.2017.08.038.

Seiler, A., S. I. Aanonsen, G. Evensen, and J. C. Rivenæs. 2010. Structural surface uncertainty modelling and updating using the ensemble Kalman filter. *SPE Journal* 15 (4): 1062–1076. https://doi.org/10.2118/125352-PA.

Siraj, M. M., P. M. J. Van den Hof, and J. D. Jansen. 2015. Risk management in oil reservoir waterflooding under economic uncertainty. In *2015 54th IEEE conference on decision and control (CDC)*, 7542–7547. IEEE. https://doi.org/10.1109/CDC.2015.7403410.

Skjervheim, J.-A., G. Evensen, S. I. Aanonsen, B. O. Ruud, and T. A. Johansen. 2007. Incorporating 4D seismic data in reservoir simulatuion models using ensemble Kalman filter. *SPE Journal* 12 (3): 282–292. https://doi.org/10.2118/95789-PA.

Skjervheim, J.-A., G. Evensen, J. Hove, and J. G. Vabø. 2011. An ensemble smoother for assisted history matching. *SPE Conference Paper*. https://doi.org/10.2118/141929-MS.

Skjervheim, J. A., R. G. Hanea, and G. Evensen. 2015. Fast Model Update coupled to an ensemble based closed loop reservoir management. In *EAGE: Petroleum Geostatistics*, 5. EAGE. https://doi.org/10.3997/2214-4609.201413629.

Stentoft, L. 2004. Assessing the least squares Monte-Carlo approach to American option valuation. *Review of Derivatives Research* 7: 129–168. https://doi.org/10.1023/B:REDR.0000031176.24759.e6.

Stordal, A. S., S. P. Szklarz, and O. Leeuwenburgh. 2016. A theoretical look at ensemble-based optimization in reservoir management. *Mathematical Geosciences* 48 (4): 399–417. https://doi.org/10.1007/s11004-015-9598-6.

Sutton, R. S., and A. G. Barto. 2018. *Reinforcement Learning: An Introduction*. Adaptive Computation and Machine Learning series, 2nd ed. The MIT Press. ISBN 9780262039246. https://mitpress.mit.edu/9780262039246/reinforcement-learning/.

Thomas, P., and R. B. Bratvold. 2015. A real options approach to the gas blowdown decision. *In Paper presented at the SPE annual technical conference and exhibition, Houston, Texas, USA, Sept* 2015. DOI: https://doi.org/10.2118/174868-MS.

Essen, G. M., M. J. Zandvlit, P. M. J. V. Hof, O. H. Bosgra, and J. D. Jansen. 2009. Robust waterflooding optimization of multiple geological scenarios. *SPE Journal* 14 (1): 202–210. https://doi.org/10.2118/102913-PA.

Van Leeuwen, P. J., and G. Evensen. 1996. Data assimilation and inverse methods in terms of a probabilistic formulation. *Monthly Weather Review* 124: 2898–2913. DOI: https://doi.org/10.1175/1520-0493(1996)124<2898:DAAIMI>2.0.CO;2.

Vossepoel, F. C., G. Evensen, and P. J. Van Leeuwen. 2025. Adaptive correlation- and distance-based localization for iterative ensemble smoothers in a coupled nonlinear multiscale model. *Mon. Weather Rev.*, 153. https://doi.org/10.1175/MWR-D-24-0269.1.

Wang, C., G. Li, and A. C. Reynolds. 2009. Production optimization in closed-loop reservoir management. *SPE Journal* 14 (3): 506–523. https://doi.org/10.2118/109805-PA.

Wang, L., and D. S. Oliver. 2019. Efficient optimization of well drilling sequence with learned heuristics. *SPE Journal* 24 (5): 2111–2134. https://doi.org/10.2118/195640-PA.

Wang, L., and D. S. Oliver. 2021. Fast robust optimization using bias correction applied to the mean model. *Computers & Geosciences* 25: 475–501. https://doi.org/10.1007/s10596-020-10017-y.

Whitaker, J. S., and T. M. Hamill. 2002. Ensemble data assimilation without perturbed observations. *Monthly Weather Review* 130: 1913–1924. https://doi.org/10.1175/1520-0493(2002)130<1913:EDAWPO>2.0.CO;2.

Willigers, B. J., S. H. Begg, and R. Bratvold. 2011. Valuation of swing contracts by least-squares Monte Carlo simulation. *SPE Economics & Management* 3: 215–225. https://doi.org/10.2118/133044-PA.

Xu, C., P. A. Dowd, K. V. Mardia, and R. J. Fowell. 2006. A flexible true plurigaussian code for spatial facies simulations. *Computers & Geosciences* 32 (10): 1629–1645. https://doi.org/10.1016/j.cageo.2006.03.002.

Zhang, Y., and D. S. Oliver. 2010. Improving the ensemble estimate of the Kalman gain by bootstrap sampling. *Mathematical Geosciences* 42: 327–345. https://doi.org/10.1007/s11004-010-9267-8.

Zhao, Y., A. C. Reynolds, and G. Li. 2008. Generating facies maps by assimilating production data and seismic data with the ensemble Kalman filter. *SPE Improved Oil Recovery Conference*. https://doi.org/10.2118/113990-MS.

Zou, C., Q. Zhao, G. Zhang, and B. Xiong. 2016. Energy revolution: From a fossil energy ERA to a new energy ERA. *Natural Gas Industry B* 3 (1): 1–11. https://doi.org/10.1016/j.ngib.2016.02.001.

Index

G. Evensen et al., *Ensemble History Matching*,
https://doi.org/10.1007/978-3-031-99155-4

GPSR Compliance

The European Union's (EU) General Product Safety Regulation (GPSR) is a set of rules that requires consumer products to be safe and our obligations to ensure this.

If you have any concerns about our products, you can contact us on ProductSafety@springernature.com

In case Publisher is established outside the EU, the EU authorized representative is:

Springer Nature Customer Service Center GmbH
Europaplatz 3
69115 Heidelberg, Germany

Batch number: 10844819

Printed by Printforce, the Netherlands